TELEPEN
002281

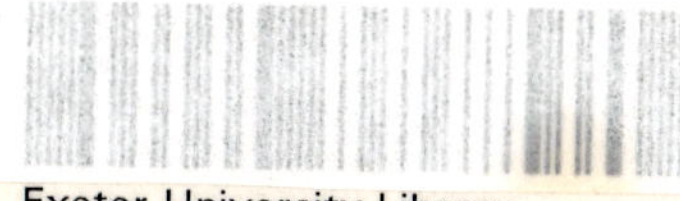
3202846475
Exeter University Library

Manganese Nodules: Dimensions and Perspectives

Natural Resources Forum Library

Volume 2

By: The United Nations Ocean Economics and Technology Office
Published for: The United Nations by D. Reidel Publishing Company,
Dordrecht, Holland / Boston, U.S.A. / London, England

Manganese Nodules: Dimensions and Perspectives

Prepared by the United Nations
Ocean Economics and Technology Office

D. Reidel Publishing Company
Dordrecht : Holland / Boston : U.S.A. / London : England

Library of Congress Cataloging in Publication Data

Main entry under title:

Manganese nodules.

(Natural resources forum library ; 2)
Bibliography: p.
Includes index.
1. Manganese nodules—Congresses. I. United Nations.
Ocean Economics and Technology Office. II. Series.
TN490.M3M33 333.8′5 78-27625
ISBN 90-277-0500-3
ISBN 90-277-0902-5 pbk.

Printed in The Netherlands.

CONTENTS

PREFACE

With the publication of the United Nations study entitled *The Future of World Economy* (1977), the United Nations has enlarged its ongoing activities to examine long-range social and economic trends in the world economy and the ways and means of influencing them in order to meet development objectives accepted by the international community.

Research and analysis have been initiated within the Dept. of Int. Economic and Social Affairs, as well as in other parts of the United Nations system, to examine alternative development scenarios for the world economy up to the year 2000. In this context, projections will be made of production and consumption in specific sectors of the world economy. Within the framework of this exercise, considerable importance will be attached to the production and the consumption of certain primary metals, including copper and nickel, and therefore, to the adequacy of these resources, which constitutes one of the basic conditions for growth. However, current estimates of the global availability of these resources have focused almost exclusively on land-based sources and generally have not taken into account manganese nodules which, until a few years ago, were considered an improbable source of metals.[1]

Given the recent technological advances for exploiting the sea-bed and the production in the near future of copper, nickel, cobalt, and possibly, manganese from sea-bed nodules, the omission of sea-bed nodules from estimates of the available stock of mineral resources is becoming increasingly untenable.

Therefore, in order to improve the accuracy of these estimates and thereby the usefulness of the economic models and projections of the future of the world economy now being established, it was considered appropriate to examine the problems of assessing sea-bed resources and to determine the manner in which estimates of these resources should be introduced into the existing estimates of the world's resource base. In view of the great uncertainties surrounding the issue and the divergent views expressed on the question in the literature, the United Nations convened a meeting of a group of high level experts who were selected from among the specialists in universities, scientific institutions, and commercial interests engaged in sea-bed mineral development. The primary purpose of the meeting was to bring to light the basic elements that need to be considered in making any estimate of the availability of this resource in the future. Specifically, the Group of Experts was called together in an effort to:

1. isolate those factors which account for the large variability among estimates of sea-bed resources and future production volumes;
2. establish statistical confidence limits for certain key parameters;
3. determine the technical characteristics of the future nodule industry;
4. determine what basic data are necessary for future work in estimating sea-bed resources and anticipating production volumes.

The meeting of the Group of Experts was held in New York from 28 November to 1 December 1977 under the auspices of the Ocean Economics and Technology Office of the Department of Economic and Social Affairs of the United Nations. In addition to a number of United Nations specialists, the following experts invited from outside the United Nations system participated:

Jagdish C. Agarwal, Director—Development, Ledgemont Laboratory of Kennecott Copper Corp., U.S.A.

Alan A. Archer, Assistant Director, Institute of Geological Sciences, London, United Kingdom.

Hubert Bastien-Thiry, Centre National pour l'Exploitation des Oceans (CNEXO), France.

Michael J. Cruickshank, United States Geological Survey, Reston, Virginia, U.S.A.

Franz F. Diederich, Research Institute for International Techno-Economic Co-operation, Technical University of Aachen, Federal Republic of Germany.

Marne A. Dubs, Director, Ocean Resources Department, Kennecott Copper, Corp., U.S.A.

Jane Frazer, Scripps Institution of Oceanography, Geological Research Division, University of California at San Diego, California, U.S.A.

Menelaos D. Hassialis, Henry Krumb Professor of Mining, Columbia University, New York. U.S.A.

Robert A. Healing, Institute of Geological Sciences, London, United Kingdom.

Alexander F. Holser, Senior Technical Advisor, Ocean Mining Administration, Office of the Secretary, United States Department of the Interior, U.S.A.

Dieter Kuchen, Research Institute for International Techno-Economic Cooperation, Technical University of Aachen, Federal Republic of Germany.

Eberhard Muller, Director, Metallgesellschaft, A.G., Federal Republic of Germany.

Hokuichiro Ohmachi, Chief, Marine Geology Department, Geological Survey of Japan, Japan.

Francisco Orrego, Director, Institute of International Studies, University of Chile, Chile.

David Pasho, Head, Offshore Minerals Section, Energy Mines and Resources, Canada.

Richard Tinsley, Second Vice-President, Continental Bank, Chicago, Illinois, U.S.A.

V. K. S. Varadan, Director-General, Geological Survey of India, India.

Conrad Welling, Manager, Ocean Systems of Lockheed Missiles and Space Company, Inc., U.S.A.

W. C. Woodmansee, International Data and Analysis, U.S. Bureau of Mines, U.S.A.

The Secretary-General of the United Nations would like to express his deepest appreciation to all members of the Group of Experts, particularly to Mr. Alan A. Archer of the Institute of Geological Sciences, United Kingdom, who, as Chairman of the Group, conducted the meeting most admirably. While by its very nature, the task undertaken by the Group of Experts is not amenable to simplistic answers, the value of the meeting stemmed from its objective review of most of the technical detrimants of deep-sea mining. The conclusions reached will be of substantial benefit to the United Nations and to all interested parties in their continuing work on assessing the potential of sea-bed mineral resources. The equitable sharing of these resources, whose development will certainly become a reality in the near future, should allow the entire international community to benefit from what has been declared by the United Nations General Assembly to be the "common heritage of mankind."

PART I

INTRODUCTION

Of the mineral resources of the deep sea-bed, the only major occurrences likely to be exploited in the immediate future are the ferromanganese nodule deposits found in all major oceans and in many different types of sedimentary environments. Since they were first recovered during the expedition of the H.M.S. Challenger (1872-1876) on 18 February 1873, 160 miles southwest of the island of Ferro in the Canary Island Group, they have been the subject of much speculation.

Initially, the topic which dominated much of the discussion on nodules was how they were formed; nodules that were recovered exhibited a wide range of variation in physical characteristics such as internal structure, appearance and size. Various theories, all complex, were proposed to explain the genesis of, and the observed variance among nodules, and to date no entirely satisfactory theory for the origin and growth of manganese nodules has been produced. Following this early interest, work on nodules became sporadic. In the "Albatross" expedition between 1899 and 1900, extensive collections of nodules from the Pacific were made which led to the identification of a zone of high nodule concentration lying off the west coast of North America between latitudes 15° N and 7° N.

As a result of a Swedish deep-sea expedition following the Second World War, an extensive collection of deep-sea sediment cores was acquired. When part of this collection was subjected to geochemical investigations, a marked correlation between some of the metal values found in nodules was noted. Even though this discovery led to another era in the investigation of nodule genesis, it was not until 1965 that the first coherent hypotheses on this subject began to appear. Between 1965 and the present, the literature on manganese nodules has undergone considerable expansion both in terms of volume and in terms of areas of specialization.

Perhaps, however, the single most important development during this period has been the emergence and growth of commercial interests in exploiting nodule deposits to recover their valuable metal content. Nodules have generated commercial interest not only because of their content of nickel, copper, cobalt and manganese but also because of a variety of other factors, such as the pressures exerted by political forces. Metal values, like other characteristics of nodules, exhibit considerable variations within and between major oceans, and in some areas the higher values compare favourably with metal values in some deposits of copper and nickel currently being exploited on land. The nature of their occurrence and the environment within which they are found, however, place considerable obstacles in the way of their exploitation.

Nodule deposits are essentially surficial deposits which occur on the sea-bed. The deposits of current commercial interest are found at water depths of between 4,000 and 6,000 metres on a terrain which shows as much, if not more, variation in relief than that which occurs on continents. Additionally, they are unevenly distributed in all the major oceans and vary over short distances both in terms of metal values and areal coverage. In the recent past, efforts have been made to quantify the metal tonnages within these deposits. The variation in these estimates attests to the number and complexity of the factors which need to be accounted for in any such effort. With an incipient nodule industry readily identifiable, there is general agreement that the metals produced from these deposits will reach world markets some time during the 1980's. The extent to which they will contribute to the world resource base in the long-term future, however, is a function of the complex factors that prompted the convening of the meeting of the Group of Experts. It was recognized at the outset that given the apparent paucity of information, the meeting would not yield a definitive set of numbers characterizing these resources; therefore, general agreement among the experts on the major factors which affect estimates of the resource and on how these factors should be perceived and taken in account was the intended objective. In this respect, the expectations of the United Nations were fully met.

The agenda of the meeting partially dictated the format of this report. Part II is concerned with the basic data and with the underlying problems pertaining to the current data base. Part III concentrates on the concept of a mine-site and the types and ranges of information required to estimate their number. Part IV, which deals with commercial considerations, examines some of the investment requirements of a nodule mining concern and some of the technical difficulties which the future nodule industry will have to overcome. Part V addresses the problems associated with recovering metals from nodules, and Part VI looks at other technological issues which can affect the exploitability of these resources. In each of these chapters, the contributions of the experts[1] are preceded by a definition of the basic problems covered in the chapter and by commentary on the relationship of the contributions to their solution. This is followed by a summary of the discussion which took place and the specific findings of the meeting with respect to these problems. Part VII represents the Secretary-General's summary of the major findings of the meeting. A selected bibliography on manganese nodules is also included.

PART II
THE BASIC DATA

CHAPTER I

PROBLEM ADDRESSED AND A SUMMARY OF THE DISCUSSION

1. Problem Addressed

Before a mineral deposit can be worked or its future development planned, its size and grade (metal content) must be determined. These essential factors may be the first, but they are by no means the only factors that have to be taken into account. In this respect, manganese nodules are no different than any mineral deposit on land. However, manganese nodule deposits are unlike other mineral deposits in several important respects: e.g., for practical purposes, they can be considered as two-dimensional rather than three-dimensional ore bodies. It follows that the area occupied by a manganese nodule deposit is a measure of its size, but as the nodules may be widely scattered or closely packed on the sea floor, the average weight of nodules present in any unit area (that is their "abundance", "concentration" or "areal density") must be established in order to determine the size of the deposit. Grade is established by subjecting samples to standard analytical techniques.

The papers by Bastien-Thiry on "Sampling and Surveying Techniques" and Frazer on "The Reliability of Available Data on Element Concentrations in Sea-Floor Manganese Nodules" discussed some of the procedures used in establishing size and grade and the sources of error in them. Bastien-Thiry emphasized that with the photographic and other techniques widely used at present, abundance could be estimated at any point with an accuracy of only about ±50%. But if several samples were collected at the same station, the error might be reduced to ±25%. With spade corers, this accuracy would be improved to ±10%. After prospecting in the vincinity of French Polynesia, the French effort was redirected towards an area in the North Pacific between North American and Hawaii, known as the Clarion-Clipperton zone, (7° N to 15° N and from 120° W to 150° W). Within this zone, a large area had been sampled at geostatistically determined intervals, in order to identify an area in which more detailed work would be justified. As a result of this reconnaissance programme, it was estimated that the average abundance of the nodules in the 2.2 million square kilometres comprising that zone was 3.45 kilogrammes (wet) per square metre. The samples also indicated that the average grade was 2.25 per cent combined nickel and copper.

Frazer discussed the errors associated with published analyses of manganese nodules and concluded that for nodules with nickel plus copper in the range of 2.5 per cent to 2.7 per cent (which accounts for a higher proportion of the nodules from the Clarion-Clipperton zone than from any other similar range), the grade could be determined to within 10 percent. Although some analyses were more inaccurate, Frazer emphasized that when the average of results from many sample stations was considered, the analytical error was small compared with the other uncertainties involved in an assessment either of potential reserves or of resources. Frazer also discussed the evidence that, in some areas at least, grade and abundance were inversely related. She also outlined steps that might be taken to increase the accuracy and usefulness of the basic data.

On the basis of the further sample data now available, Healing confirmed that the distribution of nickel and copper grades in nodules was broadly consistent with the well-known relationship postulated by Lasky, that is, geometrically increasing cumulative quantities associated with arithmetically decreasing grade. This seemed to apply to all oceans even when data were included, appropriately weighted, for the clearly anomalous Clarion-Clipperton zone. The relationship, however, could not be demonstrated for grades below 0.25 per cent nickel plus copper, nor very satisfactorily for the highest grades found: also, it did not appear to hold for cobalt. In spite of these limitations he nevertheless thought that, at this stage of the assessment of resources and reserves, generally the Lasky distribution provided a useful method for summarizing and handling the data available over most of the relevant range of grades. The possibility was recognized that ultimately other distributions, Lasky-type or not, which could better fit the observed data, would be found.

Ohmachi gave an account of the work on manganese nodules that has been carried out by the Geological Survey of Japan, largely in the Central Pacific Basin. The Japanese Geological Survey had been cooperating with scientists from the countries in the South Pacific and had included representatives from South Pacific countries on some of the cruises.

1. Summary of the Discussion

Having established the importance of grade and abundance (concentration, areal density) in the evaluation of nodule deposits and having discussed some of the techniques available for determining these two parameters, the Group focused its attention on the limitations of these techniques, particularly those mentioned in Bastien-Thiry's paper.

An initial topic brought up for discussion was the purpose of sampling in the exploration programme. Specifically, it was asked whether the techniques described were designed to determine either the in-situ or the recoverable reserves. The ensuing discussion revealed the range of exploration strategies available to nodule miners depending on what constraints were considered most important in actual mining. One viable strategy would be to undertake exploration to locate the best nodule deposit on the basis of grade and abundance and subsequently to determine recoverability. For example, if topography was deemed to be a limiting factor in mining recovery even when grade and abundance were above predetermined minimum levels, then mapping the topography could be given higher priority in an exploration programme. In this case, the general outline of the strategy would then be to locate areas of favourable topography and identify one in which grade and abundance were acceptable.

In the discussion of grades, potentially large sources of error were identified; in the analytical technique, the method of choosing sites to obtain samples, the collection of samples themselves, and extrapolation from sampled to unsampled areas.

Regarding the first source of error, it was concluded that analytical errors were not significant for estimates of resources over large areas of the sea-bed. The Group of Experts, however, recommended that in reporting nodule grades, this should be done on a dry weight basis with a report of the moisture content of the nodule. It was also suggested that selection of nonrepresentative nodule samples could introduce greater errors in average grades than those arising from inaccurate analysis, and a number of important points were raised about extrapolation errors. It was concluded that despite appreciable variation in grade over short distance, the variation was considerably reduced when these values were averaged over large areas. Therefore, the data available in the public domain were adequate for the purposes of resource assessment, but were not sufficient for the evaluation of specific mine-sites. This limitation particularly applied to the use of Lasky or other types of mathematical functions to describe the grade-tonnage relationship. Although the difficulties encountered in developing such functions were recognized, notably those concerned with the size and randomness of the sample data available and the assumption that grade varies independently of abundance, it was concluded that these did not necessarily invalidate the method for purposes of resource assessment. However, the technique had no relevance to the determination of specific mine-sites.

On the subject of nodule abundance, the general conclusion was that little confidence could be placed in published data. It was stated that current knowledge of the frequency distribution of abundance values was very imperfect. Several experts discussed the methods of collecting the data used to arrive at figures on abundance which included spade corers, dredges, grab samplers, box corers, surface and remotely operated still and movie cameras, photography and real-time television systems. Photography by itself was deemed an inadequate method of measuring abundance since not all nodules were visible; estimates based on photography alone, therefore, tended to be biased on the low side. Furthermore, shape, which was important, could not be determined from photographs. Grab samplers were also felt to be biased on the low side, while costlier box corers were much more accurate. It was reported that new techniques were being developed for the purpose of measuring abundance.

There was considerable discussion on the apparent inverse relationship between abundance and grade. In some areas, notably one area that had been identified as a potential mine-site, this relationship did not appear to exist. It was felt that more work must be done to determine where the inverse relationship held and its full implications for mining and resource inventory work. Other considerations in the relationship between grade and abundance appeared to be related to small-scale topotgraphy which would vary in importance from site to site.

The Group of Experts concluded that the factors directly related to the data that accounted for significant differences in estimates of the total resources and of the number of first generation mine-sites were as follows:

1. Data for each of the world's oceans were unevenly distributed.
2. Only a small part of the data was in the public domain.
3. The purpose for which data have been gathered varied.
4. Improvements in the tools and methods for collecting data have undergone considerable improvements in the course of time, resulting in variation in the reliability of data collected at different times.

5. In general, while grade estimates were reliable, abundance distribution was more difficult to determine.
6. As far as the results of sampling were concerned, high metal values by themselves were of little practical significance, and additional information, including data on grade, abundance distribution and topography, were required for sizable areas.
7. Minability was determined by several factors not the least important of which were grade, abundance and topography. The individual factors were more complex then they appeared at first sight and were interconnected. An integrated view of all the factors, therefore, was necessary for purposes of mine-site evaluation.

CHAPTER II

SAMPLING AND SURVEYING TECHNIQUES

1. Introduction

In the last twelve years, the subject of nodules has taken hold of people's minds and fired their imagination. Today, various views are being expressed and many research papers are being published on polymetallic nodules. The mystery surrounding this subject and the passionate reactions aroused are still growing. Underneath all this is the fact that we have here a new problem, not yet fully grasped scientifically, technologically or economically. The unknown with which we are still grappling in the area of nodules, at present, precludes us from drawing any conclusions which are too definitive.

It is true that a considerable amount of very precious and precise data and information obtained at great expense by groups or consortia remains proprietary and is not available to everyone. If all this data and information were made public for analysis, we would immediately be in a position to increase considerably the wealth of knowledge on the subject, if not to answer all the questions raised by the people interested in nodules. This is a stage that we must pass through in order to better understand, interpret and draw conclusions.

Having made these preliminary remarks, I would like to explain my approach to the subject. I used the results and findings from the first phase of the exploration effort which has been carried out for several years by the French Group AFERNOD (Association Francaise pour l'Etude et la Recherche des Nodules). I will define the scope of this work so as to present to you, as objectively as possible, the conclusions this Group has reached thus far. Table 1 gives some indication as to the time and money spent with respect to this exploration work.

2. Exploration Method

From 1971 to 1974 we carried out our first exploration work at sea with a limited budget and the idea of making an inventory of the areas close to French Polynesia. These years of "apprenticeship" led us to the following conclusions at the end of 1974:

- although certain promising occurrences did exist within the sector explored in Polynesia, the Clarion-Clipperton fracture zone appeared much richer in immediate prospects.
- the exploration equipment we used, i.e. primarily free-fall devices, appeared sufficiently rugged, economical and generally suitable for use in virtually all meteorological conditions of the Pacific.

In addition, in view of the general situation prevailing at the time, it was decided to delineate as quickly as possible an area for which a claim for exclusive mining rights could be filed. In the summer of 1975, we began an important exploration programme devoted exclusively to the Clarion-Clipperton zone.

2.1 Organization

Although the literature on nodules contained rather profuse information on the Clarion-Clipperton zone in terms of nodule *grade*, very little data were available with respect to nodule concentrations on the ocean floor. After much reflection and discussion, we decided to proceed as follows:

First phase—to cover as much of the promising area as possible, with a widely spaced but regular grid pattern.

Second phase—depending on the results obtained in the first phase, to increase the number of measurements only on those parts of the region explored at wide grid, so as to improve the accuracy of the data of interest, and hence to increase the reliability of our information.

Third phase—to modify the efforts on the basis of the information acquired and in conformity with economic considerations and requirements of the nodule recovery systems designers.

We did not seem to have estimated the potential reserves of the Clarion-Clipperton zone accurately. As it turned out, in order to select a favourable area we had been forced to survey a very large portion of the region. That being the case, we decided simply to choose a sector on which we would concentrate. If today we are satisfied with this choice, this does not mean that we would not have been just as content if our choice had gone to another sector.

2.2 Large Grid Operations

As already mentioned, the primary concern of AFERNOD at the end of 1974 was to identify as quickly as possible an area in the Clarion-Clipperton in order to file an exclusive claim. It was then a matter of determining how to proceed technically.

Two approaches were possible. The first was to do a minimum of systematic surveying and to concentrate, as soon as an occurrence appeared sufficient, all our investigations in that area in order to arrive at the delineation of a deposit of adequate extent and characteristics to constitute a mine. Or, conversely, we could perform a systematic squaring of a large portion of the promising zone and then, on the basis of the findings, concentrate on the area appearing the most favourable.

The second approach was finally selected, and from the summer of 1975, a systematic survey was begun. The results are given in tabular form in Figure 1. This represents a rigorous squaring, each apex of the chosen squares corresponding to a "locality", which taken altogether totalled approximately 300 localities.

The distance between adjacent localities, i.e. the length of each side of the unit square, was 50 nautical miles (93 km). That value was chosen on the basis of the results of a first geostatistical study which we performed by using the variogram method. From the results we had at that time, in addition to information in the relevant literature, we came to the conclusion that if, over a distance of approximately 100 km, nodules grade remained relatively constant (i.e. within two adjacent localities thus investigated, assays of the corresponding samples give the same values in copper, nickel, cobalt), there was every chance of that value remaining the same along the profile from one locality to the next. In each locality we performed two types of work: accurate measurements; and more qualitative determinations, for lack of equipment capable of obtaining very exact figures.

FIGURE 1. Exploration on a Large-Grid Basis.

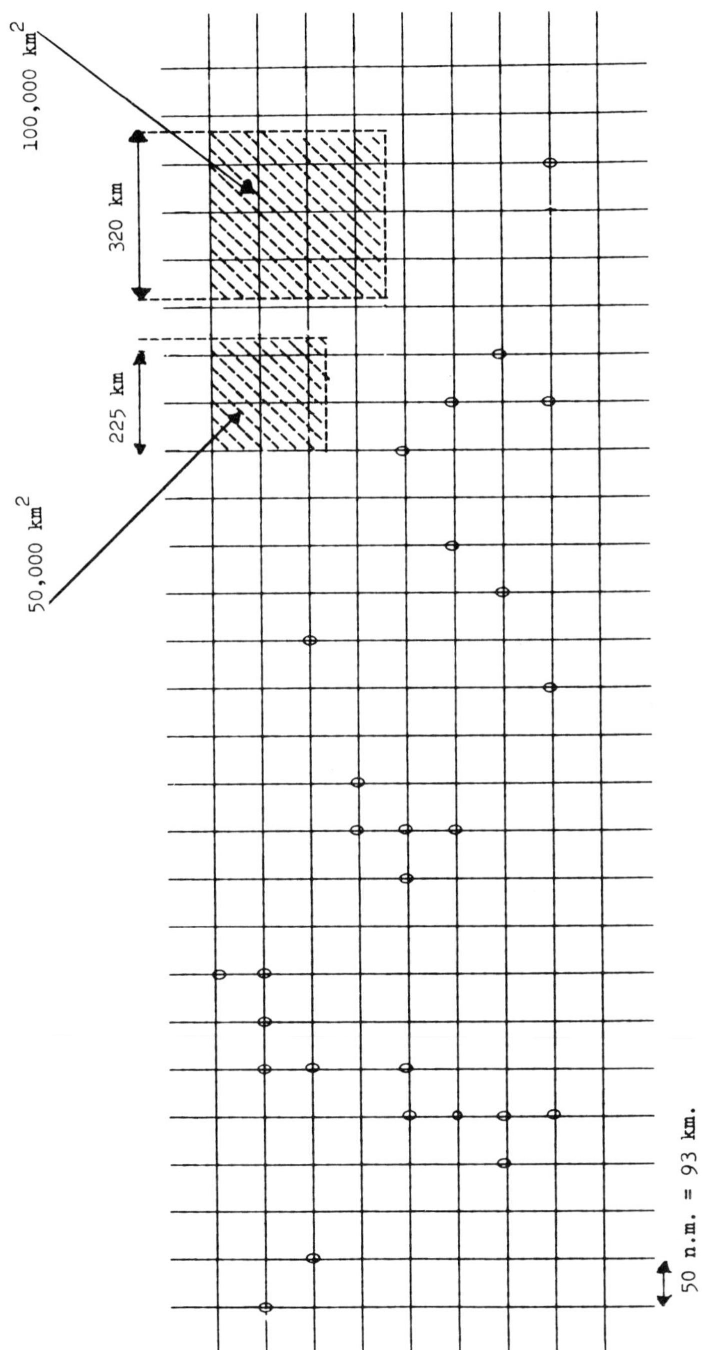

Accurate measures consist primarily of measurements characterizing the nodule ore, i.e. its content in terms of economically interesting elements, and its abundance on the ocean floor; in addition, the bathymetric surveys performed during these cruises allowed us to determine the average water depth of the areas explored; finally, a certain number of systematic measurements were made to characterize the nodules "environment".

Qualitative measurements provide a rough evaluation of the underwater relief (rough because the surface echo-sounder did not have a very high power of resolution at water depths of 5000 metres, [cf. Figure 2]). We also obtained general information on the meterological conditions of that region, which, when the time comes to proceed to commercial exploitation, will naturally be a very important economic factor.

FIGURE 2.

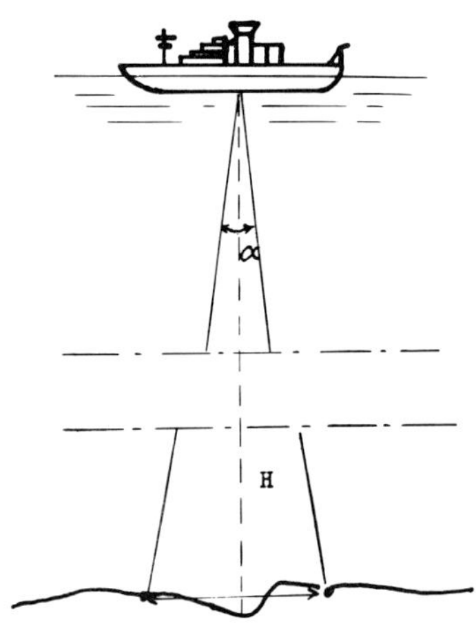

Width L, swept by an emission beam of angle x at a depth H is approximately equal to:

$H\alpha$; in radians

if, $H = 5\,000$ m and $\alpha = 10°$

$$L = \frac{5\,000 \times 10 \times \pi}{180} \cong 800 \text{ metres}$$

The main results of the wide grid exploration operations are summarized below (the ocean test had a duration of approximately 27 days of which 14 to 17 days were in the working area):

Number of ocean tests	8
Date	August 75 to September 76
Type of vessel	CORIOLIS (6), Then NOROIT (2)
Number of localities	260
Number of samples	1,560
Number of photographs	1,560
Total area covered	2,200,000 km^2

2.3 Determination of Nodule Parameters

For the present paper I shall exclude those factors which are relevant to the nodule "environment", for example measurements concerned with the sediments, the topography, the oceanological or meterological parameters. I shall deal with only two characteristics on the basis of which we chose the sector considered the most favourable within the area covered by our wide grid exploration.

It is important to say a few words about the actual operations carried out at each of the localities in the network characterized above. At each node of the grid we dropped seven combined devices in line at intervals of approximately 200 metres. As shown in Figure 3, each of these was fitted with a sample collector and a still camera. The area sampled with this type of device is approximately $0.2m^2$. The area covered by a photograph is approximately $1m^2$.

We shall suppose that the nodules recovered, and the nodules photographed are representative of the same lot. The values shown at each locality both with respect to metal content and concentration on the ocean floor, represent the *average value* for each of these parameters measured at each of the 7 sampling points.

2.3.1 Metal content of the nodules

I don't think it is necessary to discuss at length the methods used to determine this fundamental parameter. These methods do not raise any controversy. Whether analysis is performed by atomic absorption or by X-ray fluorescence, with shipboard equipment or in land-based laboratories, the results obtained are accurate and reliable. Of course, the necessary precautions must be taken to make sure of the quality and representativeness of the sample analysed; these precautions have, in my opinion, been taken by our exploration survey teams.

2.3.2 Sea-floor nodule concentration

The determination of this second parameter, however, requires further discussion and elucidation.

In view of the equipment we had, i.e. a still camera and sample collector, two methods could be used to achieve the desired result. We could assess the proportion of the area covered by nodules in the photograph and deduce their weight per unit area; or weigh the mass of nodules contained in each basket and deduce the concentration.

(A) Proportion of the area covered by nodules on a photograph. This is a classical problem of interpretation of photographs. In the case we are interested in, two main sources of error are possible. In so far as the exact estimation of the proportion of the area covered by nodules is concerned, it can be said that although it is difficult and very expensive to systematically use an analyser capable of evaluating this parameter rigorously, the estimate made by a naked but expert eye and the cross-checking with estimates performed by other experienced viewers gives a satisfactory result; the margin of error does not exceed 10%.

However, the accuracy obtained in extrapolating the nodule population per unit area from the proportion of a photographed area covered by nodules is much lower.

FIGURE 3.

Recovery Unit with radio
beacon, flash and pennant
Flotation Unit (4 balls)
Clamping device
ERJI Camera
30°
Camera starter
A: 0.45 section
B: Recovery section
A
stainless steel shaft
B
Inertial starting device
EGMO
Sampler
27 Kg ballast
ENGIN ED1

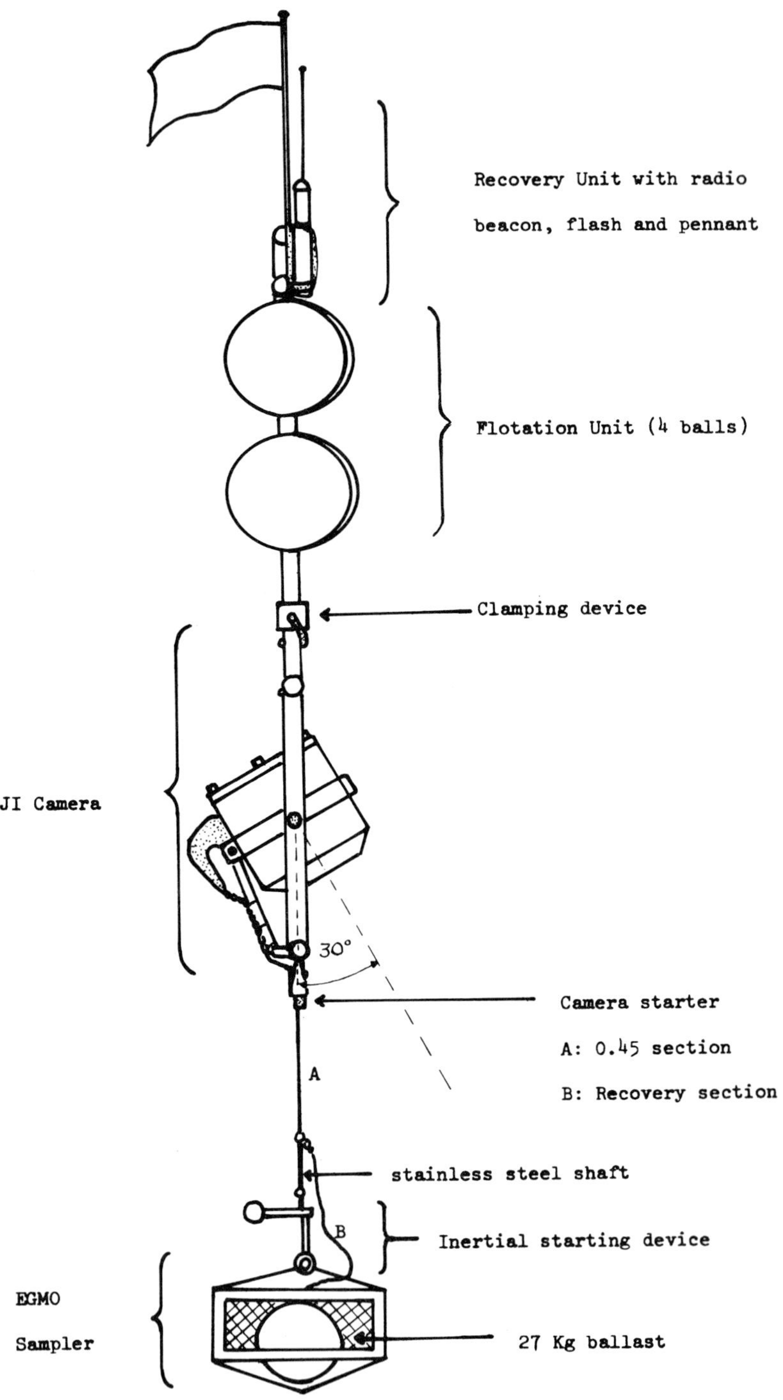

This, in fact, involves taking a two-dimensional value and from it, determining a three-dimensional one. To accomplish this, a conversion factor must be known with precision; this is the shape of the nodules.

If we suppose that the nodules are spheroidal in shape (cf. Figure 4a) this shape factor is as follows: $1/6\ d^3$, d being the diameter of the sphere.

If, on the contrary, the nodules are ellipsoidal (cf. Figure 4b) the shape factor = $1/6$ abc. Combining, for purposes of simplification, each nodule or each size fraction of nodules to a sphere having a diameter of a, the error committed on the shape factor becomes: $bc/a^2 = 3/8 = 0.375$ assuming, which is a probably practical case, that $b = 3a/4$ and $c = a/2$.

FIGURE 4a.

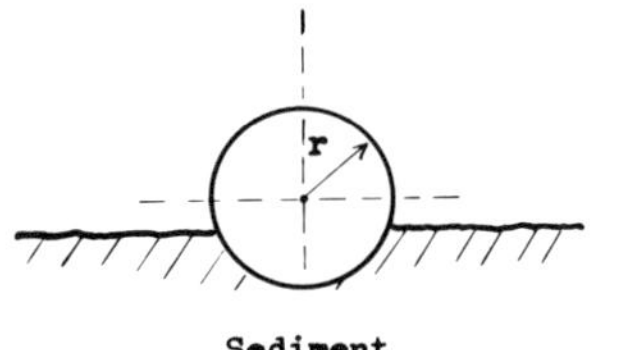

Apparent surface area : πr^2
Real volume : $4/3\ \pi r^3$

FIGURE 4b.

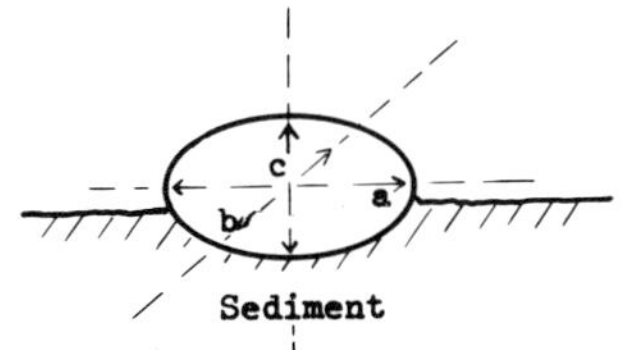

Apparent surface area : $1/4\ \pi ab$
Real volume : $1/6\ \pi abc$

(B) Amount of nodules collected: it is simple to weigh the contents of each basket, and then, knowing the theoretical area of the sample, calculate the concentration of nodules occurring on the ocean floor.

The determination is easy, but inaccurate experience shows that the sampling equipment has a nasty tendency of not picking up all the nodules in the area traversed. Furthermore, this tendency is not consistent. The pick-up rate depends on various parameters which are virtually impossible to anticipate in each case.

To remedy this situation, the tendency is to use an *average correction factor* but this can only be arbitrary, and the overall result, therefore, very inexact.

(C) In view of these considerations, our geologists developed a method for determining this parameter (nodule concentration). This involves the simultaneous use of photographs and samples taken from the same spot.

Although apparently simple in theory, in practice it is actually quite arduous, even fastidious, to apply this method station after station.

It can be expressed by the formula: C = (4/300) × S × d × f × $\sqrt{r}$ where:

C = concentration stated in wet kilograms/square metre;
S = population in % of the area covered by the nodules at the sampled spot
d = average diameter of the nodules expressed in mm
f = average flatness of the nodules
r = average roundness of the nodules

This relationship is in fact, the same as that in paragraph (A) above; it supposes, furthermore, that the density of the nodules = 2g/cm².

We have indicated above the accuracy with which we assume we know the parameter: S = 10%. It is also important to discuss the accuracy with which we can determine the other three parameters in this formula.

2.3.3 The average diameter of the nodules

To determine this parameter, we use the formula:

$$\frac{\Sigma \left[\frac{d_{i1} + d_{i2}}{2}\right] \sigma_i}{\Sigma \sigma_i}$$

in which d_{i2} and d_{i1} are the diameter fractions measured on each photograph and $\sigma_{\hat{n}}$ i the percentage of each fraction, as shown on the following table:

Diameter fractions (in mm)		%
d_{i1}	d_{i2}	
5	10	10
10	20	15
20	30	50
30	40	25

In the example corresponding to the above table: i = 4

$$d = \frac{\frac{(5+10)}{2}10 + \frac{(10+20)}{2}15 + \frac{(20+30)}{2}50 + \frac{(30+40)}{2}25}{100}$$

d = 24.25 mm

To evaluate the diameter fraction and their percentages, a square grid having the same overall dimensions as that of the photograph is used. It is obvious that this approach requires a certain amount of experience on the part of the observer, but it has been shown that this is readily acquired. An additional precaution consisted of having the same photograph interpreted separately by two or more viewers, and then comparing the results.

2.3.4 Nodules flatness

We use the following formula to determine this parameter:

$$f = \frac{\Sigma_k Y_k V_k}{\Sigma_k V_k}$$

in which y_k is the flatness corresponding to a given appearance category and v_k the corresponding proportion (%) of nodules in that category, as shown in the table below.

2.3.5 Roundness of the nodules

We use the following formula to determine this parameter:

$$r = \frac{\Sigma_e \, r_e w_e}{\Sigma_e w_e}$$

in which r_e is the roundness corresponding to a given physical appearance and w_e the corresponding proportion (%) of nodules falling into that category, as shown in the table below.

In fact, given the experimental nature of these two parameters (roundness and flatness), then: $k = e$ and $v_k = w_e$

Roundness (1st figure) Flatness (2nd figure) (in tenths)		%
7	7	15
5	3	20
3	9	25
3	5	40

$$f = \frac{7 \times 15 \; + \; 3 \times 20 \; + \; 9 \times 25 \; + \; 5 \times 40}{100} = 5.9 \text{ tenths}$$

$$r = \frac{7 \times 15 \; + \; 5 \times 20 \; + \; 3 \times 25 \; + \; 3 \times 40}{100} = 4.00 \text{ tenths}$$

The roundness and flatness of the nodules are measured either by direct measurement or by comparing the nodules recovered with the charts found in Appendix I. The percentages correspond to the weight of the respective nodules in each of the roundness/flatness categories.

It is important to know with what accuracy the nodule concentration is obtained, and this is not easy to achieve. We consider that the errors on each of the above three parameters are:

$$\frac{\Delta f}{f} \cong \frac{\Delta r}{r} \cong \frac{1}{7} = 0.14$$

$$\frac{\Delta d}{d} \cong \frac{5}{30} = 0.17$$

Under these conditions and taking into account the relative error estimated for parameter S:

$$\frac{\Delta C}{C}=\frac{\Delta S}{S}+\frac{\Delta f}{f}+\frac{\Delta d}{d}+\frac{1}{2}\frac{\Delta r}{r}=0.48$$

the relative error would thus be on the order of 50%.

Furthermore, experience has proved this to be true. It may happen that photographs of certain areas do not give evidence of the presence of nodules, although nodules have been recovered by the sample collectors. This is because in that region the nodules were beneath a thin veneer of sediment. We must then be capable of providing an ocean floor concentration figure from "sampler" data alone. In view of this situation, and taking into account the statistics we had available as a result of several hundred observations, we devised a correction factor (greater than 1) allowing us to obtain the concentration from the quantity of nodules recovered (cf. (B) above).

In reality, such a factor presents wide fluctuations. With respect to an individual sample, the correction factor may give a much greater error than that resulting from the general formula indicated in (C) above.

Last, we checked the figure calculated from the formula with those obtained by a method where the apparent diameter is determined by direct measurement of the nodules, after size classification of each sample collected. This allowed us to increase our confidence in the accuracy of the determinations.

Up to now, we consider that our method, however long, arduous, painstaking and repetitive it may be, leads to a concentration figure having a *maximum relative error of 50% in most cases.*

I have just explained how we conducted our large grid exploration operations. I described the methodology we used to determine the nodule parameters, particularly the most delicate one, i.e. nodule concentration on the sea-floor. From this effort, and from the data arrived at, we have drawn two types of conclusions: We have estimated the total tonnage of recoverable metals in the 2.2 million square kilometre zone systematically surveyed at large grid. Above all, we have selected, within this area, the sector appearing to us to be the most favourable for the concentration of all our work at sea since September 1976.

3. Extrapolation from the Results

3.1 Total Tonnage in the Clarion-Clipperton Zone

From the measurements obtained with respect to nodule grade and concentration at each of the 300 localities visited, we propose to evaluate the total nickel and copper tonnage of nodules lying in the 2.2 million square kilometre area between the Clarion-Clipperton fractures.

This we would do by multiplying the average global values obtained for these 300 localities regarding both nodule grade and concentration by the total area of the zone.

But to legitimise the use of such a simple arithmetical operation, it is necessary to muster supporting evidence, for example, by treating the whole mass of data collected, by geostatistical methods. This is precisely what our geologists did; they amply manipulated the variograms established on these two parameters, and their examination supported the legitimacy of using the elementary calculation above.

In short, having arrived at an average value of 3.45 km/m² for the 300 localities surveyed, representing approximately 2000 samples, we deduced that the tonnage of nodules (wet) contained in the area of some 2.2 million square kilometres is:

$$\underset{\text{tons/square metre}}{3.45 \times 10^{3}} \times \underset{\text{square metre}}{2.2 \times 10^{6} \times 10^{6}} = 7.6 \times 10^{9} \text{ tonnes}$$

Based on the average Cu and Ni assays obtained and assuming a metallurgical recovery of 90%, this tonnage of nodule represents:

66×10^{6} tons of Ni

58×10^{6} tons of Cu.

If, in addition, we are only interested in exploiting a nodule deposit having the following characteristics:

Ni + Cu = 2.5%

Minimum concentration = 10 kg/m²,

the the above total tonnage drops to 0.35×10^{9} tons or 22 times less; if we wish to mine these 350 million tons over a 20-year period, and assuming a recovery efficiency of the nodule collecting system of 20%, we then have reserves capable of producing an annual:

30,000 tons of Ni

25,000 tons of Cu

or virtually the totality of the nickel requirements of a country like France and 5 to 7% of its copper requirements.

3.2 Value of Such Estimates

Two factors must be taken into consideration: (1) the accuracy of the basic parameters, particularly as concerns nodule concentration on the ocean floor; and (2) the representativeness of the whole conglomerate of measurements with respect to the totality of the area considered.

As to the accuracy of the concentration measured, we may recall the considerations developed above as to the methods employed to calculate this parameter. We know that despite all the precautions and all the care taken, in the final analysis the nodule concentration figure obtained may be as much as 50% off.

Regarding the representativeness of the measurements, 2000 measurements were made. Each one covered roughly 1m². We have thus operated on a cover of 2×10^{3} km².

In other words, the total area of the region considered is 10^{9} (a billion times) larger than the area surveyed. Such a ratio is, naturally, mind boggling! Our faith must therefore be very great in terms of both the continuity of the phenomena generated by nature and of our mastery of the geostatistical tools, in order to give sufficient credibility to the bold extrapolations made.

Just how far should our credulity go? This is not easy to quantify. The considerations developed in the following section will contribute to strengthening our confidence.

4. Verification of Results: Assessment of a Potential Deposit

Figure 1 shows that a number of localities in the grid explored are marked with the symbol 0. These are points where the required combination of assay plus concentration, as defined in section 3.1 above, are indeed satisfied. (Actually, the concentration figure accepted is 7.5 kg/m^2.)

Within this sub-zone, consisting in reality of several sectors, we augmented the number of stations through 9 additional campaigns thus increasing the density of our data in a ratio of close to 8.

The data were obtained either at close grid of 2 km or with a medium sized grid pattern of 30 km.

We have not yet finished analysing this mass of data, but we can already see that the *new average concentration* figure is higher than what was obtained with a wide grid pattern.

Such results are obviously very encouraging.

This result can also be considered as not very representative since the ratio of 1 billion indicated above has after all only been reduced by one power of 10. Furthermore, it is obviously not materially possible to continue along the present lines; several generations of explorers would not suffice! It is therefore necessary to proceed in a different manner.

We thus wish to operate in two stages. The first state, to commence in 1978, will involve continuous observation of the ocean floor along a certain number of profiles, using a photographic process capable of taking a great number of pictures at short intervals. Our main aim here is to check nodule concentration ratios to make sure they are at least as high as those obtained previously.

It should be noted that other nodule explorers, including Deep Sea Ventures and the German Group AMR, used this method of observation very early in the game, particularly by employing a towed T.V. camera operating in "real-time".

The second and final stage will involve mapping, for the benefit of the dredging operation, the distribution of nodules on the sea-floor and limiting areas inaccessible to the nodule collector.

How much detail should this map show? What techniques could be used to produce this map? It is too early to give a final answer to this question. The design concepts and basic equipment already exist, but the combination in which these should be used will gradually become clearer as more knowledge is gained, and as the day approaches when nodules are lifted to the surface and metals are extracted.

5. Conclusion

In the preceding sections, I have dealt at length with the manner in which the French Group AFERNOD has conducted its exploration efforts to date. From the mass of experimental data obtained from small areas, we extrapolated the total potential orebody, without, however, losing sight of the uncertainty involved in such an ambitious assessment. We are fully aware that more sophisticated techniques of observation and measurement than those used in the work carried out to date will have to be developed.

I shall conclude with this important remark; the main purpose of our exploratory work was not to determine the actual number of mine-sites having the required characteristics, but through comparative measures to choose the sector within the zone investigated constituting the most attractive potential orebody. This we believe has been accomplished even if much further work still remains to be done before this area can legitimately be considered a true nickel, copper, cobalt and manganese deposit.

CHAPTER III

THE RELIABILITY OF AVAILABLE DATA ON ELEMENT CONCENTRATIONS IN SEA-FLOOR MANGANESE NODULES

1. Introduction

Since the mineral resources of the deep ocean have been declared "the common heritage of mankind", estimating the magnitude of these resources is of concern to all peoples. For such estimates to be credible all interested parties must be able to evaluate them independently. This obviously requires a large reliable public data base. The data preferably should also come from a number of independent sources to ensure the absence of systematic error. Of course, when one compiles a data base from a wide variety of sources, it is very difficult, if not impossible, to ensure consistent quality. This paper will address the problems of evaluating present data on element concentrations in manganese nodules and determining what can be done to improve the quality and consistency of such data.

To ascertain the reliability of publicly available data on element concentrations, we must consider several factors. First, how accurate and precise are the analyses themselves? Next, even if the analyses are accurate, how well do they represent the actual element concentrations in the sea-floor deposits? Finally, even if the available analyses are truly representative of the locations from which they were collected, how adequate is the data base for predicting element concentrations in nodules throughout the world ocean?

After discussing the reliability of available data I will recommend some procedures to improve the usefulness of our future data base. These suggestions will not provide definitive solutions to all the problems, but I hope they will serve as points for discussion.

2. The Data Base

The public data base consists of all nodule analyses reported in the published literature or in unpublished theses, reports of oceanographic institutions, etc. It also includes analyses by various academic, government or industrial laboratories that have been made available but are not formally reported anywhere.

In 1972 the U.S. National Science Foundation, Office for International Decade of Ocean Exploration (IDOE) sponsored compilations of the available data[1,2] which then included nodule analyses for 530 stations. A 1976 compilation by Monget et al[3] contains 1270 analyses of marine manganese nodules, of which almost 1100 are identified as to location from which the samples were collected. These data are stored in a data bank at the Center of Geological Information, Ecole Nationale Superieure des Mines de Paris. Since this data bank is kept up to date, it presumably contains more analyses now. I do not know the search capabilities of this system, nor whether information is available to the public on request.

Since 1968 we have maintained at Scripps Institution of Oceanography (SIO) an ocean sediment data bank which includes manganese nodule analyses. Our nodule data base presently contains almost 3800 analyses for about 1850 sampling sites. Some 350 unpublished nodule analyses from our own laboratory, performed by Mary Fisk and the author, are included. Other information stored in the data bank includes locations of 55,000 stations, with complete sediment descriptions for many of these. Where such information is available we have included indications of the presence or absence of manganese nodules, percentage of nodule coverage of sea-floor photos, and nodule concentration in kg/m^2. We update our files about six times per year, and we believe it is the largest and most accessible collection of nodule data currently in the public domain.

3. Analytical Precision and Accuracy

The methods most widely used today for bulk elemental analyses of nodules are atomic absorption spectrometry and x-ray fluorescence spectrometry (energy dispersion and wavelength dispersion). These methods are much easier and quicker for large numbers of samples than old-fashioned wet chemistry, but they still rely ultimately on the analysis of standards by chemical separation procedures and therefore are not more accurate.

Each method has its advantages and disadvantages, but if properly performed, can provide precise and accurate results. While preparing this paper I consulted many scientists from industry and academia, and they all agreed on one point—all trust their own laboratory's analyses but regard others as dubious. I believe most analysts, most of the time, obtain results adequate for estimation of sea-bed mineral resources, and I will present evidence below to support this viewpoint. We are not interested, after all, in results with a precision of parts per million. Errors of 10 to 15% of the amount present for an individual analysis are of little consequence when the value we use for predictions is the average of a number of analyses. The public data base probably includes some analyses with very large errors; but since reported values are likely to be normally distributed around the correct value, they will average out and have little influence on our estimates unless they occur in analyses of isolated samples. Considering the magnitude of uncertainty in other parameters that enter into a resource estimate, we should not focus too much on our somewhat inaccurate data about nodule grade.

The precision of an analysis is a measure of its reproducibility by the same technique. Imprecision, which is due to such factors as counting statistics or the imprecise weighing or measuring of substances, is random, and will have a normal distribution about a mean value. An analyst can produce results with an arbitrary precision by repeating the experiment a number of times. The average statistical errors (one sigma) reported by analysts from a number of different laboratories are about .03% each for copper or nickel, .75% for manganese, .3% for iron, and .004% for cobalt in manganese nodules. The typical statistical error on a value for copper plus nickel, therefore, would be .043%.

The accuracy of an analysis indicates how close the result is to the "true" concentration of the element in the sample. Inaccuracy is frequently expressed as "systematic error," because such errors tend to trend in one direction for a given analyst using a particular method. When we combine analyses from a number of sources, however, these errors also tend to be random.

Inaccuracy arises from several causes. It must be kept in mind that none of these techniques we are discussing measures the absolute quantity of anything. The unknown sample is compared to one or a series of calibration standards, and we know the amount of an element in the sample only in relation to the amount present in our standard. Accurate analyses thus depend upon an accurate knowledge of what is in the standard.

An important characteristic of most techniques used in nodule analysis is that the apparent amount of an element being measured is greatly affected by the amount and nature of the other elements in the sample (the so-called "matrix effect"). A high-iron, low-manganese nodule might have exactly the same nickel content as a low-iron, high-manganese nodule, but it would appear different in the raw analytical data. Either the sample must be diluted to minimize matrix effects, or corrections must be applied to compensate for them. While corrections are usually based on physical theories, they are in fact quite empirical. The development of reliable empirical correction procedures requires considerable experimentation with a variety of samples covering the concentration range expected in the unknowns.

Nodules fresh from the ocean contain a large amount of water, and some of the water evaporates gradually as the nodules are exposed to air. To give a standard basis for comparison of nodule compositions, analytical results are usually reported for samples oven-dried at 100-110° C until no further water evaporates. This removes water amounting to about 30% of the original sample. Since nodules easily reabsorb 10-15% of their weight of water from the atmosphere, especially after being powdered, samples dried before preparation must be redried before analysis. If nodules are analysed without drying the element concentrations reported will be low, as compared to nodules treated by the standard procedure, and if they are dried at higher temperatures some tightly bound water will be removed and results will be high. Nonstandard drying procedures lead to inconsistencies in the data base, but they are not inaccuracies per se.

Analysing metal oxides in nodules is a relatively new game. While the literature is full of papers on techniques for analysing silicate minerals or metal alloys, little has been published on the analysis of manganese nodules. That difficulties exist in nodule analysis is illustrated by one mining company's report on materials they use as calibration standards. This report, prepared in 1973, contains individual analyses of each standard by a number of laboratories. For comparison we will consider only the results for materials which contain the elements of interest in about the same range, 0.5-1.5% copper or nickel. The reported standard deviation on the copper analyses for several non-nodule samples is 4-5% of the amounts present, whereas the standard deviation for nodules is 12 to 15% of the amounts present. For nickel, standard deviations are 5% of the amount present for a sulfide, 18 to 33% for manganese nodules. Thus we must conclude that as recently as 1973, at least some laboratories could not analyse nodules with an accuracy approaching that of their analyses for other ore materials.

Fortunately, most of the public data have been generated since then. One way to get an idea of current analytical accuracy is to examine the results of a round-robin exercise conducted recently by Kennecott in cooperation with the National Science Foundation International Decade of Ocean Exploration Manganese Nodule Project. A homogeneous powdered sample of nodule material was prepared by Werner Raab of Kennecott and splits were sent to a number of different laboratories. The results as reported by 19 different analysts showed the following:

(i) Some analysts dried the samples at temperatures around 110° C before analyzing, but almost two-thirds of them did not. Although the standard drying procedure removed water amounting to about 10 % of the sample, the average concentrations of all elements for samples analyzed as received were only 6 % lower than those for the samples that were dried before analysis. This effect is easy to explain for X-ray fluorescence analyses because water has a much lower mass absorption coefficient than other sample constituents and thus absorbs X-rays less. Theoretically a similar effect should not occur with atomic absorption: a 10 % difference in sample mass should produce a corresponding 10% difference in analytical results. Why it did not is unclear. It is true that nodules on the sea floor contain much more water, but I know of no laboratories that analyze fully saturated nodules. Much of the water evaporates on exposure to air, and I doubt that many analyzed samples contain more than 10 to 15 % water that would be removed by the standard drying procedures.

If the round-robin is representative of analytical practice in general, we might conclude that more than half the reported analyses are of samples with some unknown water content. The effect of this is to lower the reported average grade for a region by only four or five %. On the other hand, this may be compensated for by analyses of samples dried at higher temperatures.

(ii) One analyst reported a cobalt content which differed from all others by at least a factor of four; even with this value omitted, the reported cobalt content ranged from a 0.11 to 0.24%, which covers 80% of the range for nodules from the northeastern equatorial Pacific. From this we must conclude that while we probably have a good idea of average cobalt value in a well-sampled area, we have no real knowledge of variations in cobalt within the area if data are from several laboratories. For data from a single analyst the relative variations in cobalt it contains may be reported, but the reported average concentration may be innacurate.

(iii) The analyses are all on a dried basis. We find the standard deviations for both nickel and copper are each about 0.09 %, 9 % of the amount of nickel present in the sample and 13 % of the amount of copper. The standard deviation on manganese is 0.72 %, (about 3 % of the amount present) and on iron 0.56 % (about 5 % of the amount present). These standard deviations, of course, include statistical as well as systematic errors. Because errors add in quadrature the standard deviation on nickel-plus-copper is 0.13 % (about 8 % of the amount present).

Since we can expect that 95 % of the analyses will be within two standard deviations of the correct value, we can estimate that errors in the measurements of copper and nickel will rarely exceed ±0.26 % for a similar nodule. We would, of course, expect the errors in the manganese content of the nodules to vary somewhat according to the amount of these elements present in the sample analyzed, but I have no other sets of inter-laboratory comparisons with which to calculate expected errors over the whole range of nodule compositions. For nodules in the range of 2.5 to

2.7 % nickel plus copper, the range in which nodules most frequently occur in the area between the Clarion and Clipperton Fracture Zones, we can expect almost all copper-nickel grade data to be accurate to within about 10 % of the amount present.

Just how significant are errors of up to 10% in individual analyses? Not very, when we remember that these errors are essentially random when analyses from more than one laboratory are averaged together and that random errors add in quadrature. On an average grade for five samples the error would be about 5% of the amount present; on 25 samples, about 2%, and on 100 samples, less than 1%. We whould probably add in an estimated error of about -5% due to analysis of undried samples. Even so, analytical errors are quite small in comparison to uncertainties in other factors that enter into a resource prediction.

4. Representativeness of Analyses

We must now consider how well the material selected for analysis represents the bulk composition of a nodule, and further, how well a single nodule represents the element concentration for a composite of all the nodules that might be retrieved from a single location on the sea-floor.

The sample which best represents the composition of a whole nodule is, of course, the whole nodule itself. Investigators do not always analyse a whole nodule, however. Part of the reason for this is that large nodules take a large amount of time to grind and homogenize, but the main reason is that we wish to save as much nodule material as possible for other studies. Sample curators from academic institutions are generally reluctant to provide any more material than is absolutely necessary for a project. We generally prepare a half or a quarter nodule unless the nodule is a centimetre or less in diameter, and a study in our laboratory indicates that the imprecision introduced by this procedure is only about 4% of the combined nickel and copper. When a cross-sectional segment amounting to only 1/16 of the total nodule is analysed, we find the error thus introduced increased to 9%. Since the Mn/Ni and Mn/Cu ratios for such analyses of nodule cross-sections remain almost constant within the limits of analytical precision, we have concluded that the errors introduced are mainly due to the inclusion of varying amounts of detrital material within the different nodule segments.

To be representative of the bulk sample, a nodule section must contain representative portions of all the different parts of a nodule. It is well known that the various elements are not distributed uniformly throughout a nodule. Raab[4] has shown that the outer rind of nodules is different from the interior and that nickel and copper are concentrated in that part of the rind in contact with the sediment, whereas they are depleted in part of the rind in contact with seawater. Marchig and Gundlach[5] reported a series of nodule analyses indicating that nickel is enriched in the outer 1-5 mm layer of nodules, whereas copper is enriched in the interior. Work by the author and Fisk at Scripps Institution of Oceanography confirms the effect noted by Marchig and Gundlach and suggests that this effect may account for a significant part of the compositional variation in nodules from the same location, where nodules of different size and shape have varying fractions of their total bulk in the nickel-rich outer layer.

Many but not all nodules contain an observable nucleus; this may be some foreign material or a nodule fragment which differs only slightly in composition from the rest of the nodule. The nucleus may make up a large fraction of the bulk of the nodule or may be insignificant in size. Thus a representative nodule segment must include appropriate proportions of the nucleus and other layers.

The question that immediately presents itself is whether we should exclude from our data base all analyses of nonrepresentative nodule parts. I think not, because predicting sea-floor mineral resources is only one of several legitimate reasons for studying manganese nodules, and a central data bank serves a number of different purposes. Such analyses can be excluded from studies aimed at determination of sea-bed resources. On the other hand, even if analyses of outer rinds or nodules minus their nuclei differ from analyses of the whole nodules, they can give us an upper limit for nickel content and probably a fair estimate of copper content.

More work needs to be done to enable us to use these analyses of partial nodules correctly. In the meantime, it is interesting to note the results of a number of studies we have made at SIO with the information in our data bank. We very frequently calculate the average grade for an area two ways—with all analyses included, and with only those we know to be representative bulk analyses. We have yet to find a case in which the difference is significant.

The analysis of a single nodule, no matter how accurate, does not necessarily indicate the average composition of all nodules from that site on the sea-floor. (By "single site" I mean the area that would be sampled by a box core or grab sampler.) Several investigators have reported that compositional variations among nodules from the same site may be greater than variations among different sites within the same region. At SIO we recently analysed between six and sixty separate nodules from each of three box cores collected from within 75 km of 14° 45'N, 125° 30'W. Water depth at the three sites was between 4339 and 4651 m. The average copper plus nickel content for nodules from each of the three sites varied only between 2.50 and 2.56%, practically identical within analytical precision. Within a single box core, however, nodule grade varied from 1.97 to 2.91% nickel plus copper.

Forty samples from a dredge haul analysed by Glasby[6] ranged from .39 to 1.38% copper plus nickel, about the same spread as for our box cores, although a much greater percentage variation. We would expect a wider spread for a dredge haul, which samples a much larger area than a box core.

Samples from a grab analysed by Marchig and Gundlach showed a nickel plus copper range from 1.86%, for a nodule with a large nucleus of volcanic material, to 3.48%. This box grab and another dredge haul analysed by Glasby had bimodal distributions, with very few of the nodules having a grade anywhere near the average for the station.

Thus it seems that the selection of nonrepresentative nodules for analysis could introduce errors in estimating average nodule grade much larger than those introduced by inaccurate analyses. Where a number of analyses are available for an area, however, these errors probably cancel each other as do the other errors.

Even with the wide variations we have mentioned, we can still tell something from a single analysis about whether the nodule grade for a given site is above or below required mine grade. In a check through data on 171 stations for which several analyses are available, we found only two stations at which the average grade for Ni + Cu was above 2.25% (the required average grade for first generation mining) and one of the

an idea of current analytical accuracy is to examine the results of a round-robin exercise conducted recently by Kennecott in cooperation with the National Science Foundation, International Decade of Ocean Exploration Manganese Nodule Project.

A homogeneous powdered sample of nodule materials was prepared by Werner Raab of Kennecott, and splits were sent to a number of different laboratories. The results, as reported by 19 different analysts, showed the following:

(i) Some analysts dried the samples at temperatures around 110° C before analysing, but almost two-thirds of them did not. Although the standard drying procedure removed water amounting to about 10% of the sample, the average concentrations of all elements for samples analysed as received were only about 6% lower than those for the samples that were dried before analysis. This effect is easy to explain for x-ray fluroescence analyses because water has a much lower mass absorption coefficient than other sample constituents and thus absorbs fewer x-rays. Theoretically a similar effect should not occur with atomic absorption; a 10% difference in sample mass should produce a corresponding 10% difference in analytical results. Why this did not is unclear.

It is true that nodules on the sea-floor contain much more water, but I know of no laboratories that analyse fully saturated nodules. Much of the water evaporates upon exposure to air, and I doubt that many samples analysed contain more than 10 or 15% water, which would be removed by the standard drying procedures.

If the round-robin is representative of analytical practice in general, we might conclude that more than half the reported analyses are of samples with some unknown water content. The effect of this is to lower the reported average grade for a region by only about 4 or 5%. On the other hand, this may be compensated for by analyses of samples dried at higher temperatures.

(ii) One analyst reported a cobalt content which differed from all others by at least a factor of four. Even with this value omitted, the reported cobalt content ranged from .11 to .24%, which covers 80% of the range for nodules from the northeastern Equatorial Pacific. From this we must conclude that while we probably have a good idea of average cobalt value in a well-sampled area, we have no real knowledge of variations in cobalt within the area if data are from several laboratories. For data from a single analyst, the relative variations in cobalt content may be reported reliably, but the reported average concentration may be inaccurate.

(iii) After normalizing the analyses so that all are on a dried basis, we find the standard deviations for nickel and copper are each about .09%, 9% of the amount of nickel present in the sample and 13% of the amount of copper. The standard deviation on manganese is .72% (about 3% of the amount present), and on iron .56% (about 5% of the amount present). These standard deviations, of course, include stastistical as well as systematic errors. Because errors add in quadrature, the standard deviation on nickel plus copper is .13%, about 8% of the amount present.

Since we can expect that 95% of the analyses will be within two standard deviations of the correct value, we can estimate that errors in the measurement of copper plus nickel will rarely exceed .26% for a similar nodule. We would, of course, expect the magnitude of the errors to vary somewhat according to the amount of these elements present in the sample analyzed, but I have no other sets of inter-laboratory comparisons with which to calculate expected errors over the whole range of nodule compo-

analyses showed a grade below 1.75% (the mining cutoff grade); we also found only two sites at which a single analysis was above 2.25% but the average below 1.75%.

5. Comparison of the Public Data With a Large Consistent Data Base

All the error estimates given above were developed for this meeting by a survey of data immediately available. There has been no time to do further analyses or perform experiments to check the validity of my conclusions. It might be desirable to have a close look at the questions of grade variations within a nodule deposit and errors introduced by nonrepresentative sampling. I hope that these matters will be discussed later.*

As one check on my findings, let us compare the public data base for the region between the Clarion and Clipperton Fracture Zones with a much larger data base that contains analyses of representative samples from evenly distributed stations performed in a consistent manner. This is the data base that has resulted from the exploration of the French Group.** Within this region the French Group has analysed nodules from more than 1800 stations as compared to less than 300 in the public data base. We cannot look at the actual data of the French Group because it is not available, but we can look at the results reported recently by Bastien-Thiry et al.[7]

These authors reported that half the weight of nodules they collected from this region had a combined nickel-copper content of more than 2.5%. Fifty-two per cent of our analyses are above this grade. Average grade was 2.25%; for our data it is 2.34% if we include only analyses representative of bulk nodules, 2.35% if we include all analyses. These results illustrate the point I made previously that when numerous analyses are averaged together, nonrepresentative sampling tends to cancel out.

We had expected our data to show a slightly lower grade because we have included samples analysed with varying water content. This seems to be more than compensated for by a sampling bias toward the richer areas. Much recent sampling (by Hawaii Institute of Geophysics and National Oceanic Atmospheric Administration, for example) for which we have analyses has been concentrated in areas already known to contain nodules rich in copper and nickel.

Figure 1 shows the frequency distribution of copper and nickel in nodules from between the Clarion and Clipperton Fracture Zones as presented in Bastien-Thiry et al[7], and superimposed on it is the frequency distribution for approximately the same area from the SIO data bank. The only real difference is that ours is more spread out, with values both above and below those reported by the French Group. This is what we would expect, since the French Group investigators took care to analyse representative samples at each station, while our data sources have not always done so. The most frequent grade range for the French data is 2.4 to 2.6%, and for our data it is about the same, 2.5 to 2.7%. Average cobalt content as determined by the French Group is .24%; our average is .22% for bulk data and .21% for all data.

Can the differences between number of potential mine-sites in this area predicted by Bastien-Thiry et al and the various numbers estimated from the public data base be accounted for by differences in estimated nodule grades between the two data bases? The

*See Chapter II.1.
**See Chapter IV.3.

FIGURE 1.

METAL GRADE IN NODULES
FROM BETWEEN
CLARION AND CLIPPERTON
FRACTURE ZONE
S.I.O. sediment data bank, Sept. 1977
Bastien - Thiry, et. al., 1977
% of analyses
30
20
10
0.4
1.0
2.0
3.0
3.6
% Ni + Cu

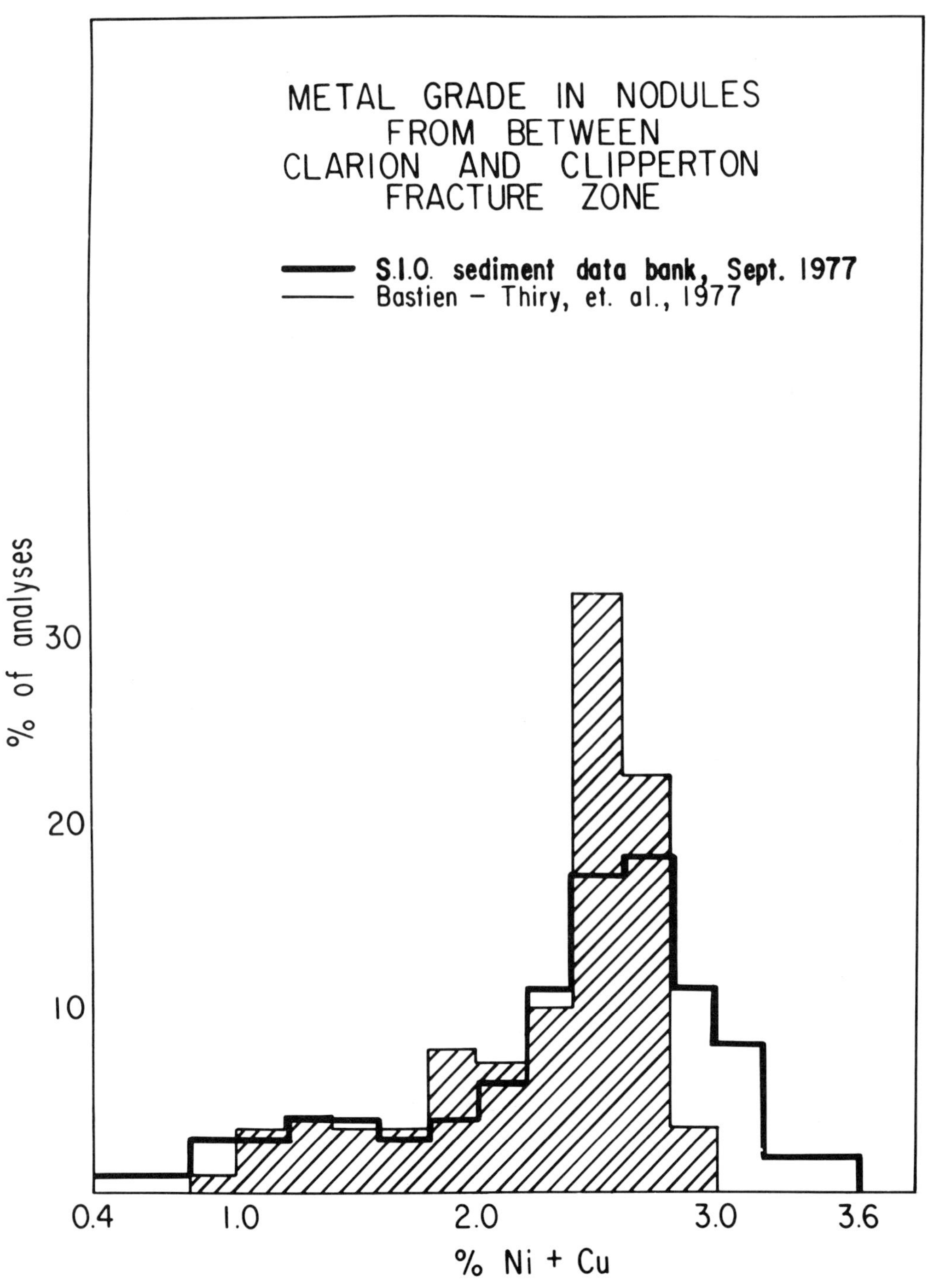

answer is clearly negative. The widely varying estimates of the number of mine-sites must be attributed to varying estimates of nodule abundance and, even more importantly, to different methods of interpreting the given data.

6. The Relationship Between Nodule Grade and Abundance and its Effect on Resource Prediction

We can rely on the public collection of miscellaneous analyses to predict average nodule grade and frequency distribution of element concentrations for a well sampled region because, although nodule concentration is very patchy and shows large variations over small distances, nodule grade is relatively constant over large areas (Bastien-Thiry, Siapno, private communications). Nevertheless, we must be aware of a type of local variation in nodule grade that may be quite significant for estimating the magnitude of nodule resources.

The regions where manganese nodules are rich in copper and nickel are characterized by rolling abyssal hills. Moore and Heath[8] surveyed a small area and found nodules preferentially concentrated on the slopes. Margolis and Burns[9] noted from surveys of several sites between the Clarion and Clipperton Fracture Zones that nodules are more abundant on hilltops and slopes than at the base of hills or in the valleys between. Skornyakova and Zenkevich[10] have made a detailed study of several localities in the Pacific Ocean and also find that nodules are more abundant on local topographic highs. Skornyakova[11] further found a higher concentration of iron in nodules from hilltops, possibly connected to underwater weathering of volcanic rocks, which leads to a decrease in the copper and nickel content of the nodules. Thus a high abundance of nodules is associated with lower nodule grade.

Menard[12] found separate evidence of an inverse relationship between grade and abundance from an analysis of all available Pacific data as a function of geological age. As crustal age increased from 20 to 120 million years, the percentage of sea-floor photographs showing nodules increased from 10% to 47%. At the same time, the percentage of samples with 1% or more nickel decreased from 33% to zero. As Menard and I studied the data further, we found additional evidence that grade and abundance are negatively correlated, and our results are in preparation.

From Bastien-Thiry's plot of resources as a function of cutoff grade we see that the curves representing nodule grade for concentrations less than 10 kg/m^2 (wet) are quite different from the curve for concentrations greater than or equal to 10 kg/m^2 (wet). It seems that nodules with grades up to almost 3% were found in areas with the lower concentrations; but where concentration is greater than or equal to 10 kg/m^2 (wet), the highest grade is about 2.65%.

On the other hand, Piper et al[13] obtained different results in their recent study of a micro area centred at about 15° N, 126° W (within the area selected as a mine-site by Deep Sea Ventures, Inc.). They found nodule concentrations to be greater in the valleys than on the hilltops. From their analyses we see no correlation between nodule grade and concentration.

More work must be done to establish where the inverse correlation between nodule grade and abundance occurs and to determine its full implications for sea-floor mining. But where it does prevail, this effect can cause us to overestimate nodule resources in an area. Let us consider how this can occur.

Skornyakova and Zenkevich present the results of a detailed survey for an area about 20 x 30 km located at about 11° N, 153° W. Samples were collected from water depths ranging from 4735 to 5070 m. We can see quite clearly from Figure 2 that nodule grade and concentration are negatively correlated here. Water depth and concentration are also negatively correlated. Average nodule concentration for an area is 8.66 kg/m^2 (wet), and average grade is 1.53% copper plus nickel. If we are not aware of any association between grade and abundance, we might reasonably decide to calculate the amount of copper and nickel that occurs in nodules here by multiplying the average grade times average concentration. From this we would estimate that nodules in this area contain about 92 tons of these metals per km^2. Suppose, on the other hand, that we consider each station separately and assume that it represents only the area immediately surrounding it. We then multiply grade for each station by concentration at the same station. Assuming that stations are relatively evenly distributed throughout the area, we can then average these results. In this case we find that we predict only 75 tons per km^2 of nickel and copper. Thus an assumption that grade and abundance are independent leads to a 24% overestimation of copper and nickel for this locality.

Moritani[14] has studied an area about 5 to 10° N, 170 to 175° W. From his data we find the average nodule concentration to be 5.9 kg/m^2 (wet) and the average grade, 2.06%. (In Moritani's paper the concentrations were shown only as ranges; I estimated that the concentration for each location was at the midpoint of the indicated range.) Calculating by the first method above we would estimate 86 tons copper and nickel per km^2 for this area; but assuming grade and abundance are not independent and calculating by the second method, we estimate 67 tons per km^2, 27% less.

FIGURE 2. Grade vs. Concentration for 10 sites in the North Pacific Ocean.

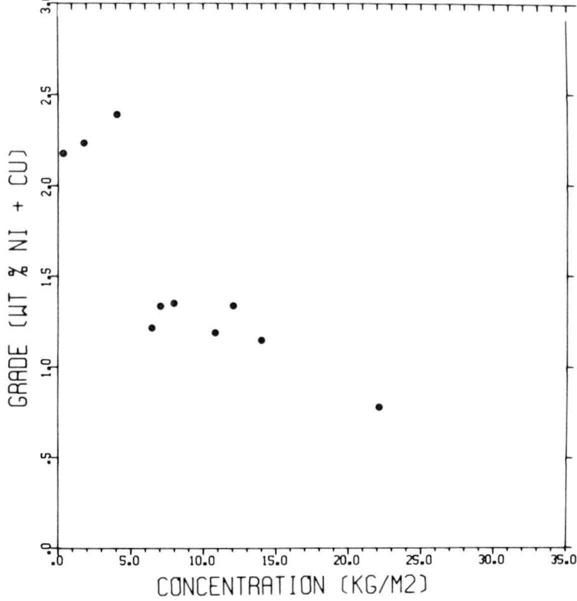

A locality in the South Pacific was also studied by Skornyakova and Zenkevich. Here nodules are much more abundant, 16 kg/m^2 (wet), while grade is quite low, 0.8%. Again, however, by the first method we estimate 92 tons nickel and copper per km^2, whereas with the second method we obtain 81 tons per km^2. Overestimation here was 14%. We should note that this area, with a grade less than half the cutoff required for mining, actually contains more copper and nickel in nodules per km^2 than the two North Pacific areas mentioned, one of which has a grade above the cutoff for mining. This points up the fact that estimating the amount of various metals in sea-floor nodules and estimating the amount of these metals that could be recovered from currently envisioned mining operations are two quite different problems.

Perhaps this negative relationship between grade and abundance partially accounts for the French Group's estimate of potential mine-sites in the northeast Equatorial Pacific being lower than those of other authors. They have detailed data on nodule concentrations for each locality in their grid, whereas the rest of us have available only scattered concentration data and are thus obliged to average them over large areas.

7. Adequacy of the Data Base

I believe we can safely conclude that, for the northeastern Equatorial Pacific, element concentration data available to the public are adequate for estimates of nodule grade. Such grade estimates are surely more accurate than estimates of available nodule tonnage based on public data. Although I have not discussed the public data on nodule concentration, I will guess that it is sufficient to allow us to predict the number of mine-sites in the northeastern Equatorial Pacific within a factor of two. To obtain data adequate for more precise predictions would require expensive exploration that is not likely to be undertaken except by potential miners, who will keep the results secret.

Data on nodule grade for the world ocean outside the northeastern Equatorial Pacific are quite sparse, with analyses for only 1300 stations—about one for every 250,000 km^2. It is some consolation to note that the frequency distribution of nodule grades within this data set is almost exactly the same overall and within each individual ocean as reported by Healing and Archer[15] for an earlier data set, which included analyses for only 530 stations. This indicates that we might be able to draw conclusions from our present limited data base that will hold up even when much more data are available. Frequency distributions for large regions, however, are more useful for estimating total resources than for predicting specific potential mining areas.

We must also remember that we have other information besides nodule analyses which can perhaps be used to help us predict nodule grade. Various investigators[16] have studied the relationships between nodule grade and their size and shape. It is even accepted by some (Pasho, Siapno, Bastien-Thiry, private communications) that one can get a good idea of the composition of a nodule just by looking at it. So we may be able to get additional grade information from sea-floor photographs.

We also have information available on sea-floor sediment types, sediment deposition rates and currents. The type of environment in which nodules are likely to be both abundant and high in nickel and copper has been fairly well established[17 18], so we can use these existing data to predict where minable nodule deposits might occur. Even if we do not wish to rely on such relationships to predict resource magnitude, we can use such information to select areas in which further exploration is most likely to prove fruitful.

Do we have enough data on element concentrations to predict resource magnitude for the world ocean outside of the northeastern Equatorial Pacific? Probably not with any degree of certainty, although I believe we have enough to eliminate some areas from consideration. The answer to this question seems to have little practical significance; decisions will be made and actions will be taken or avoided on the basis of the knowledge available and the predictions made from it, regardless of whether the data are sufficient. To explore the entire ocean even in the detail that the French Group has explored the eastern Equatorial Pacific would cost about $600 million, and it is a pretty safe bet that this is not going to happen. Intensive exploration will be carried on only where there is some probability that the value of the deposits will exceed exploration costs. Decisions on whether or where to undertake further exploration must themselves be based on available information whether it is adequate or not.

A more useful question to ask is, how best to use the available information to estimate resources. That question will be addressed by other participants.

We do want to make future additions to the data base as accurate and useful as possible. The following are my recommendations for improving the quality of data on element concentrations in manganese nodules:

(i) Results would be most consistent (but not necessarily most accurate) if a single investigator analysed all samples. This is impossible and probably undesirable. We need a number of people actively involved in studying nodules so that there will be many people capable of evaluating the analytical results. We must be concerned with the acceptability as well as the accuracy of the analyses. Consistency between analyses performed by different scientists using different methods can be greatly improved by increased exchange of samples and standards among as many investigators as possible. If different analysts are getting widely different results from the same materials, they need to find out why and correct the problems, not just gossip about the quality of each other's work.

(ii) A wider range of manganese nodule calibration standards is needed. Flanagan at the U.S. Geological Survey has prepared two natural standards and distributed samples of them to a number of laboratories for analysis. We hope the precise composition of these standards will be established soon and that they will become available for use by everyone. Still more standards are needed. Studies should also be made on the feasibility of using artificial standards.

(iii) The best information for economic evaluations comes from analysing samples that represent the total collection of nodules at a particular location: several samples or a composite which accurately represent the nodule types and size ranges found at the site. Pasho (private communication) has proposed that analysing a composite made up of three nodules selected at random from the total sample gives an adequately representative sample. This seems reasonable, although I have not had time to check it. Since my work has other aims besides estimates of nodule resources, I would prefer to have individual analyses of three nodules selected from the total sample plus their weighted average, which would presumably give the same value as analysing a composite and would provide additional information about nodule composition as related to morphology.

(iv) If, for some reason, an investigator selects a particular nontypical nodule for analysis, he should describe the size, shape and texture of the nodule and the part analysed. The entire sample collected at that site should be described as well—amounts of nodules in each size range and morphological category.

(v) We should not expect all scientists to be interested in analysing representative bulk samples. In fact, if we do only bulk analyses we may miss important clues to nodule development. It would be very helpful, however, if scientists who do choose to make special studies of parts of nodules could make their leftover nodules available to other investigators for bulk analyses.

(vi) We need more information that will enable us to relate nodule grade to concentration. If at all possible, information on nodule concentration obtained from weighing the totality of nodules collected in a grab sampler or box core, and accompanying sea-floor photographs should be made available. Persons collecting nodules for analysis should make an effort to use sampling devices that can give an indication of nodule concentration.

(vii) It is usually impossible to publish all the details mentioned above along with the results of a study, either because of a journal's page limitations or because such information would be irrelevant to the main topic of the paper. In such cases, the information should be deposited in a data bank and noted in the papers so that those who need the information can obtain it readily.

(viii) Not all effort should be restricted to measuring "tonnage and grade" of nodule deposits. Studies of the relationship between nodule chemistry and physical characteristics, factors regulating nodule composition and growth, differences between buried and surface nodules, or the location of the various elements within the nodule can have widely useful applications in the future, even though they do not help us estimate the number of first-generation mine-sites.

8. Conclusions

Estimating the magnitude of mineral resources from sea-floor manganese nodules requires a knowledge of nodule grade, nodule concentration, and detailed sea-floor topography. Our public information on nodule concentrations is limited and probably inaccurate. Bastien-Thiry reports that even with the French Group's sophisticated survey methods involving several grab samplers and sea-floor photographs at each locality, their estimated error on nodule concentration for a given locality ranges from 20 to 50%. Our data are less reliable, since they are generally based on either photographs or grab samplers for a given location, not both. The errors in nodule concentration data are certainly larger than the errors in grade data. Because nodule concentration varies more than grade on a small scale, we need even more individual data points; instead we have fewer. Our knowledge of detailed bottom topography is almost nil for most of the world ocean. Thus I conclude that although the public data we have on nodule grade are not as accurate, representative or numerous as we would like, they are not the limiting factor in predictions of sea-floor nodule resource magnitude.

There is much we can do to improve the usefulness of additions to the public data base, as I have recommended above. Efforts should also be made to correct suspected errors in the existing data. The amount of grade information could be considerably increased at little cost by the analysis of nodules already collected in the archives of various oceanographic institutions.

On the other hand, there is little likelihood that improved sampling or better analytical techniques or greatly increasing the data will lead to general agreement on the resource potential of manganese nodules if some adopt the attitude, "Our organization has lots of secret information. We can't tell you what it is, but since all the public data are unreliable you should accept our results on faith."

To facilitate international negotiations on the utlization of ocean mineral resources we need some estimate of resource magnitude and mining potential that is believed by most of the parties involved. The general acceptability of such an estimate seems almost more important than its actual precision. Thus data adequate to back up a resource prediction must be widely available. If we decide the present public data base is insufficient for this purpose, we must find a way to improve it and to ensure that adequate data are accesible to everyone.

Notes and References

1 Frazer, J.Z. and G. Arrhenius, 'World-wide Distribution of Ferromanganese Nodules and Element Concentrations in Selected Pacific Ocean Nodules', *International Decade of Ocean Exploration*, Technical Report No. 2, 1972, 51 p. (Unpublished manuscript).

2 Horn, D.R., M.N. Delach and B.M. Horn, 'Metal Content of Ferromanganese Deposits of the Oceans', *International Decade of Ocean Exploration*, Technical Report No. 3, 1973, 51 p. (Unpublished manuscript).

3 Monget, J.M., J.W. Murray and J. Mascle, 'A World-wide Compilation of Published Multicomponent Analyses of Ferromanganese Concretions', *International Decade of Ocean Exploration*, Technical Report No. 12, (NSF Grant 33616), 1976, 127 p.

4 Raab, W., 'Physical and Chemical Features of the Pacific Deep Sea Manganese Nodules and Their Implications to the Genesis of Nodules', Horn, D.R. (Editor), *Ferromanganese Deposits on the Ocean Floor*, The Office for the International Decade of Ocean Exploration, National Science Foundation, Washington, D.C., 1972, p. 31-50.

5 Marchig, V. and H. Gundlach, 'Zur Geochemie von Manganknollen aus dem Zentralpazific und Ihrer Sedimentunterlage', *Geologisches Jahrbuch*, 1976, 16: 59-77.

6 Glasby, G.P., *The Geochemistry of Manganese Nodules and Associated Pelagic Sediments from the Indian Oceans*, Ph.D. thesis (Unpublished), University of London, U.K., 1970.

7 Bastien-Thiry, H.B., J.P. Lenoble and P. Rogel, 'French Exploration Seeks to Define Mineable Nodule Tonnages on Pacific Floor', *Engineering and Mining Journal*, July 1977, p. 86-87, 171.

8 Moore, T.C. and G.R. Heath, 'Manganese Nodules, Topography and Thickness of Quaternary Sediments in the Central Pacific', *Nature*, 1966, 212: 983-985.

9 Margolis, S.V. and R.G. Burns, 'Pacific Deep-Sea Manganese Nodules: Their Distribution, Composition and Origin', *Annual Review Earth and Planetary Science Letters*, 1976, Vol. 4., p. 229-263.

10 Skornyakova, N.S. and N.S. Zenkevich, 'Regularities of the Space Distribution of Ferromanganese Nodules', Bezrukov, P. (Editor), *Ferromanganese Nodules of the Pacific Ocean*, Transactions of the P.P. Shirshov Institute of Oceanology, Moscow, 1976, Vol. 109, p. 37-76.

11 Skornyakova, N.S., 'Chemical Composition of Ferromanganese Nodules of the Pacific', Bezrukov, P. (Editor), *Ferromanganese Nodules of the Pacific Ocean*, Transactions of the P.P. Shirshov Institute of Oceanology, Moscow, 1976, Vol. 109, p. 190-240.

12 Menard, H. W., 'Time, Chance, and the Origin of Manganese Nodules', *American Scientist*, 1976, 64 (5): 519-529.

13 Piper, D.Z., W. Cannon and K. Leong, 'Composition and Abundance of Ferromanganese Nodules at DOMES Sites A, B and C: Relationship with Bathymetry and Stratigraphy', Piper, D.Z. (Compiler), *Deep Ocean Environmental Study: Geology and Geochemistry of DOMES Sites A, B and C, Equatorial North Pacific*, U.S.G.S. Open-file Report 77-778, Menlo Park, Ca., 1977.

[14] Moritani, T., 'Morphological Types of Manganese Nodules of the Central Pacific Basin', *Marine Sciences Monthly,* 1975, 8 (12): 53-58.
[15] Healing, R.A. and A.A. Archer, 'Manganese Nodules — An Example of Lasky Distribution?', *Resources Policy,* 1975, Vol. 1, No. 6, p. 306-312.
[16] Meyer, K., 'Surface Sediment and Manganese Nodule Facies Encountered on RV Valdivia Cruises', *Marine Technology,* 1973, 4 (6): 196-197.
[17] Greenslate, J.L., J.Z. Frazer and G. Arrhenius, 'Origin and Deposition of Selected Transition Elements in the Sea-bed', *Nodules in the Pacific and Prospects for Exploration,* Symposium, Valdivia Manganese Exploration Group, University of Hawaii and IDOE-NSF, 23-25 July 1973, p. 45-69.
[18] Cronan, D.S., 'Deep-Sea Nodules: Distribution and Geochemistry', Glasby, G. (Editor), *Marine Manganese Deposits,* Elsevier: Oceanography Series, Amsterdam, 1977, p. 11-44.

CHAPTER IV

THE FREQUENCY DISTRIBUTION OF NICKEL, COPPER AND COBALT GRADES IN MANGANESE NODULE DEPOSITS*

1. Introduction

In practice, a deposit of manganese nodules can be regarded as a two-dimensional orebody, as most lie either on the surface or only partly buried by sediment. Analyses of superficial samples obtained by grabs or dredges therefore may be taken to represent the grade of the whole of the orebody at the sampled point. On the basis of this assumption, the frequency with which different grades occur was examined in Healing and Archer[1], with particular respect to the extent to which the distribution of nickel and copper in nodules conformed to that suggested by Lasky. For certain mineral deposits on land, Lasky[2] postulated that the cumulative tonnage of ore available increases geometrically as the average grade decreases arithmetically. This contribution re-examines the relationships.

In 1975, data were readily available[3,4] from a total of 530 stations in all oceans: an updated compilation was published in 1976[5]. The Data Bank at the Scripps Institution of Oceanography of the University of California, San Diego, is constantly updated; the data from 1523 sations that were in this Data Bank in June, 1977, are the basis of this study. The procedure used by Healing and Archer has been followed again. For ease of comparison, the analogous tables have been given the same numbers.

The extent of the necessary assumptions and the limitations of the data must again be emphasized:

(i) Some published results almost certainly include analytical errors, or are not accurately representative of whole nodules.** However, elimination of these results would involve the introduction of unacceptable subjective judgements and it is therefore assumed that any errors are not systematic. Where there is more than one analysis for a station the average metal content is used.

(ii) The randomness of the earlier samples was questioned. For example, the stations in the Clarion-Clipperton area (defined as from 7° N to 15° N and from 120° W to 155° W) accounted for 9% of the 530 stations: although the total area in which nodules occur is not known, it has been suggested[6] that it may be approximately 54 million km^2, of which the Clarion-Clipperton area may account for about 3%*. This bias is more marked in the present sample, 15% of the stations being in the Clarion-Clipperton area. The number of stations in this area has increased by 460% whereas the remainder have increased by 270%.

(iii) The quantities of contained metàl can be evaluated if the distribution of the total tonnage of nodules according to grade is known, but published information on the abundance (that is the weight of nodules per unit area) and grade at the same station remains very sparse. Recent work by Menard and Frazer[7] points to an inverse relationship between the abundance and copper and nickel grade in the Pacific Ocean, with statistically significant but weak correlation coefficients. However, comparable information is not available for the other oceans and, indeed, much less is known of the worldwide distribution of abundance than of grade and much of the information is substantially less reliable as, *inter alia,* it must be derived from data on the nodule population (the proportion of the sea-bed area occupied by visible nodules). Therefore, to enable comparison of the two studies, the assumption has again been made that abundance varies independently of grade. On that basis, the relative frequencies of the samples found at each grade may be taken as a first approximation to the relative *in situ* tonnages of the different grades. That is to say the relative number of samples at each grade and the frequency distributions exemplify the tonnage-grade relationship.

(iv) The frequency of samples containing less than 0.25% of nickel plus copper clearly does not follow the same distribution as for higher grades (Table 1) and *these have again been excluded from further consideration.*

2. Method

The observed frequencies, that is the number of stations reporting data, at each grade were basically presented (Table 1 for combined nickel and copper) in group intervals of 0.25%. As in the Healing and Archer investigation these were then rearranged in descending order of grade (Table 2) and the frequencies cumulated (Table 3) as required by the Lasky concept. For working purposes all with a grade above 2% were included in a single group (and, as explained above, all values below 0.25% were ignored): the adoption of a different grouping would have some effect on the results.

The cumulative frequencies (T) at each average grade (G_A), calculated on the basis of the midpoint of the corresponding increment, were fitted by least squares to the logarithmic equation:

$$G_A = K_1 - K_2 \log T \qquad \text{or} \qquad \log T = \frac{K_1 - G_A}{K_2}$$

which expresses the relationship between the cumulative tonnage of ore and the average grade suggested by Lasky. Similar equations, but dependent on cutoff grade (G_C) were also fitted:

$$G_C = K_3 - K_4 \log T \qquad \text{or} \qquad \log T = \frac{K_3 - G_C}{K_4}$$

The results derived for nickel and copper combined, for both G_A and G_C, are shown in Table 4.

The curves for average grade and cutoff grade were each fitted separately from the observed data. This would not be necessary if exact functional relationships existed, in which case one could simply and theoretically be deduced from the other. Under the conditions of closely but not perfectly fitting curves, however, the above equations yield the following empirical relationship:

$$G_A = \frac{K_2}{K_4} G_C + K_1 - \frac{K_2 K_3}{K_4}$$

*One of the authors (J.Z. Frazer) gratefully acknowledges the partial support of the Scripps Institution of Oceanography by the United States Bureau of Mines (Grant No. G0264024).

**See, for example, Chapter II.3.

*See Appendix A.

TABLE 1.
Summary of Data in Scripps Data Bank (June 1977): Nickel Plus Copper

Nickel plus Copper %	Number of stations (percentage of totals in brackets)									
	Atlantic Ocean*			Indian Ocean*	South Pacific Ocean*	North Pacific Ocean			All oceans	
	North	South	All			Clarion-Clipperton	Rest	All	All	Except Clarion-Clipperton
0 —0.25	15 (16)	29 (22)	44 (20)	22 (16)	21 (5)	1 (0)	23 (5)	24 (3)	111 (7)	110 (8)
0.25—0.50	45 (47)	50 (39)	95 (42)	52 (37)	123 (29)	1 (0)	79 (16)	80 (11)	350 (23)	349 (27)
0.50—0.75	25 (26)	23 (18)	48 (21)	31 (22)	128 (30)	2 (1)	79 (16)	81 (11)	288 (19)	286 (22)
0.75—1.00	8 (8)	15 (12)	23 (10)	13 (9)	58 (13)	11 (5)	72 (14)	83 (11)	177 (12)	166 (13)
1.00—1.25	1 (1)	6 (5)	7 (3)	14 (10)	32 (7)	9 (4)	49 (10)	58 (8)	111 (7)	102 (8)
1.25—1.50	—	4 (3)	4 (2)	2 (1)	21 (5)	21 (10)	29 (6)	50 (7)	77 (5)	56 (4)
1.50—1.75	—	—	—	4 (3)	21 (5)	15 (7)	36 (7)	51 (7)	76 (5)	61 (5)
1.75—2.00	1 (1)	1 (1)	2 (1)	1 (1)	11 (3)	16 (7)	61 (12)	77 (11)	91 (6)	75 (6)
2.00—2.25	—	1 (1)	1 (0)	1 (1)	8 (2)	30 (14)	28 (6)	58 (8)	68 (4)	38 (3)
2.25—2.50	—	—	—	—	4 (1)	32 (14)	26 (5)	58 (8)	62 (4)	30 (2)
2.50—2.75	—	—	—	1 (1)	3 (1)	37 (17)	12 (2)	49 (7)	53 (3)	16 (1)
2.75—3.00	—	—	—	—	—	28 (13)	11 (2)	39 (5)	39 (3)	11 (1)
3.00—3.25	—	—	—	—	—	14 (6)	1 (0)	15 (2)	15 (1)	1 (0)
3.25—3.50	—	—	—	—	—	3 (1)	1 (0)	4 (1)	4 (0)	1 (0)
3.50—3.75	—	—	—	—	—	1 (0)	—	1 (0)	1 (0)	—
Totals	95	129	224	141	430	221	507	728	1523	1302

Note: not all percentages add up to 100% due to rounding.

*The boundaries between Indian and Atlantic Oceans are taken at 30° E, between Indian and South Pacific at 150° E and between South Pacific and Atlantic at 70° W.

TABLE 2.
Observed Frequencies at Each Grade*

Grade (%Ni + Cu)	Atlantic Ocean	Indian Ocean	S. Pacific Ocean	N. Pacific Ocean**	All Oceans **
>2.00	1	2	15	108	126
1.75— <2.00	2	1	11	64	78
1.50— <1.75	0	4	21	39	64
1.25— <1.50	4	2	21	33	60
1.00— <1.25	7	14	32	51	104
0.57— <1.00	23	13	58	74	168
0.50— <0.75	48	31	128	80	287
0.25— <0.50	95	52	123	79	349
Totals	180	119	409	528	1236

*Except below 0.25% (Ni + Cu)
**Including Clarion-Clipperton (weighted data)

TABLE 3.
Observed Cumulative Frequencies at Each Cutoff Grade
(with corresponding average grade in brackets)

Cutoff grade (%Ni + Cu)	Atlantic Ocean	Indian Ocean	S. Pacific Ocean	N. Pacific Ocean*	All Oceans *
2.00	1(2.125)	2(2.375)	15(2.292)	108(2.462)	126(2.437)
1.75	3(1.958)	3(2.208)	26(2.115)	172(2.243)	204(2.222)
1.50	3(1.958)	7(1.875)	47(1.896)	211(2.129)	268(2.080)
1.25	7(1.625)	9(1.764)	68(1.735)	244(2.026)	328(1.950)
1.00	14(1.375)	23(1.375)	100(1.540)	295(1.871)	432(1.752)
0.75	37(1.064)	36(1.194)	158(1.296)	369(1.671)	600(1.506)
0.50	85(0.816)	67(0.931)	286(0.996)	449(1.486)	887(1.222)
0.25	180(0.583)	119(0.688)	409(0.809)	528(1.319)	1236(0.983)

*Including Clarion-Clipperton (weighted data)

As the amount of the contained metals is a function of the quantity of nodules above the cutoff grade and the average grade corresponding to that cutoff, the preceding equations can be used to generate a further curve showing the cumulative proportion of the metals (M) present in nodules above any cutoff grade. The general form of the equation is:

$$M = G_A^T = \left(\frac{K_2}{K_4} G_c + K_1 - \frac{K_2 K_3}{K_4} \right) 10^{\left(\frac{K_3 - G_c}{K_4} \right)}$$

This contribution goes beyond the earlier work in examining briefly the separate frequency distributions of nickel and copper, as well as of cobalt and nickel equivalent. Similar procedures to those described above, but using smaller class intervals, enabled the equations in Table 7 to be derived from the data in Tables 8, 9, 10 and 11.

TABLE 4.
Derived Equations Relating Cumulative Frequency (T) with Average Grade G_A and Cutoff Grade G_c

Ocean	Average grade (%Ni + Cu)	Cutoff grade (%Ni + Cu)
Atlantic	$G_A = 2.231 - 0.730 \log T$	$G_c = 1.968 - 0.775 \log T$
Indian	$G_A = 2.667 - 0.950 \log T$	$G_c = 2.254 - 0.961 \log T$
S. Pacific	$G_A = 3.608 - 1.054 \log T$	$G_c = 3.487 - 1.230 \log T$
N. Pacific (c)	$G_A = 5.998 - 1.690 \log T$	$G_c = 7.568 - 2.657 \log T$
All oceans:		
(a)	$G_A = 5.231 - 1.378 \log T$	$G_c = 5.420 - 1.685 \log T$
(b)	$G_A = 6.985 - 1.846 \log T$	$G_c = 7.647 - 2.367 \log T$
(c)	$G_A = 5.700 - 1.511 \log T$	$G_c = 5.920 - 1.843 \log T$

(a) Excluding Clarion—Clipperton
(b) Including Clarion-Clipperton
(c) Including Clarion-Clipperton (weighted data)

3. The Effect of the Clarion-Clipperton Area

In the Healing-Archer investigation of possible Lasky-type grade distributions the data from stations in the Clarion-Clipperton area, which followed a markedly different pattern from those recorded elsewhere, were excluded although it probably contains the richest deposits. To exclude this area from the calculations imposes a restriction on the usefulness of the curves derived for all oceans.

This limitation has now been removed. Examination of the more extensive data now available, while confirming the atypical distribution of grades in the Clarion-Clipperton area (Figure 4), shows that their inclusion with the other data leads to the derivation of Lasky curves of good fit. (Re-examination shows that this also applies to the earlier data.) The results obtained for all oceans, on the basis of both including and excluding the Clarion-Clipperton area, are given in Tables 5 and 6 and illustrated in Figure 1.

TABLE 5.
All Oceans: Comparison of Expected and Observed Cumulative Frequencies

Cutoff grade	(a)		(b)		(c)	
(%Ni + Cu)	Expected	Observed	Expected	Observed	Expected	Observed
2.00	107	97	243	242	134	126
1.75	151	171	310	333	183	204
1.50	212	233	396	409	250	268
1.25	298	289	504	486	342	328
1.00	420	391	643	597	467	432
0.75	591	557	821	774	638	600
0.50	832	843	1046	1062	871	887
0.25	1170	1192	1335	1412	1191	1236

(a) Excluding Clarion-Clipperton. The expected frequencies are calculated from: $\log T = 5.420 - G_c/1.685$
(b) Including Clarion-Clipperton. The expected frequencies are calculated from:
$\log T = 7.647 - G_c/2.367$
(c) Including Clarion-Clipperton (weighted data). The expected frequencies are calculated from:
$\log T = 5.920 - G_c/1.843$

FIGURE 1. All Oceans: Cumulative Percentages of Nodules Against Cut-off Grade.

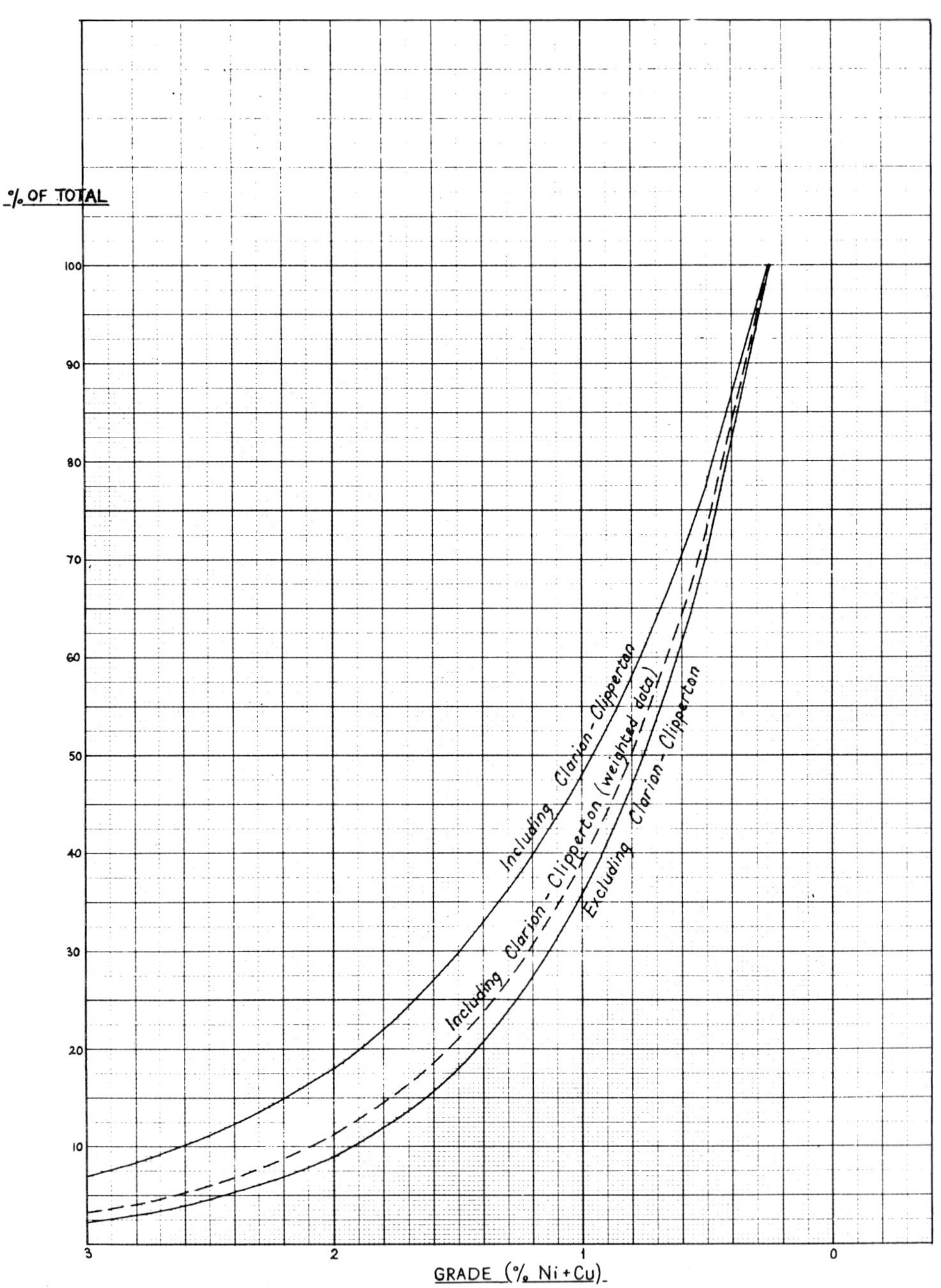

However, inclusion of all the Clarion-Clipperton data given this area a disproportionate influence. The bias can be reduced by weighting the Clarion-Clipperton data according to the proportion of the total oceanic coverage of nodules for which this area is believed to account. Although not known with any accuracy this is estimated to be about 3%*. Since the samples attributed to the Clarion-Clipperton area in Table 1 account for 15% of the total for all oceans, only one-fifth of the tabulated values at each grade need be taken as contributing in correct proportion to the all-oceans totals. The results obtained on this weighted basis are also shown in Tables 5 and 6 and Figure 1.

It can be said that the likely distribution of cumulative frequency with cutoff grade lies between the lower curve in Figure 1, which takes no account of the Clarion-Clipperton area, and the upper curve which gives it full weight. Within the envelope defined by these curves, the distribution may be more likely to be near the curve based on the inclusion of a weighted sample from the Clarion-Clipperton area than its limits. *For all subsequent calculations,* therefore, the contribution from this area to the relevant total has been reduced to one-fifth of the tabulated data. Weighting only the data from this area, rather than from all, can be justified as a first approximation on the grounds that in it the distribution is notably more anomalous and the sampling bias appears to be much greater.

4. Combined Nickel and Copper

For all oceans, the differences between the observed cumulative percentages at different cutoff grades (Table 3) and the values obtained from the equation can be seen in Table 5 and Figure 2. An analogous comparison for each of the increments in cutoff grade is shown in Table 6. The differences between the 'observed' average grades at different cumulative frequencies (or tonnages in the relationship postulated by Lasky) and the values obtained from the equation can also be seen in Figure 2, together with a curve showing the cumulative proportion of the metals present above any cutoff grade. It may be noted that similar curves generated by data excluding all from the Clarion-Clipperton area ((a) Table 4) would correspond very closely with those shown in Figure 2 in Healing and Archer. The observed and expected frequencies of both average and cutoff grades for individual oceans are shown graphically in Figure 3.

The fit between the observed and expected data for individual and all oceans remains good, although not quite as good as before for the North Pacific and all oceans (now including the weighted data from the Clarion-Clipperton area). For these, it is at the highest values that the fit is least impressive; the equations imply the existence of nodules with higher grades than those yet recorded. It is clear from the histograms (Figure 4), which show the distribution both including and excluding all Clarion-Clipperton data, that the bimodal distribution in the North Pacific contributes to the differences between the observed and expected values. A bimodal distribution is not obvious in the other oceans.

In the Clarion-Clipperton area (Table 1 and Figure 4) the combined nickel and copper grade is more than 2% at about two-thirds of the stations and above a cutoff grade of 1.75% at nearly three-quarters.

*See Appendix A.

FIGURE 2. All Oceans: Cumulative Percentages of Nodules and Metals Against Grade

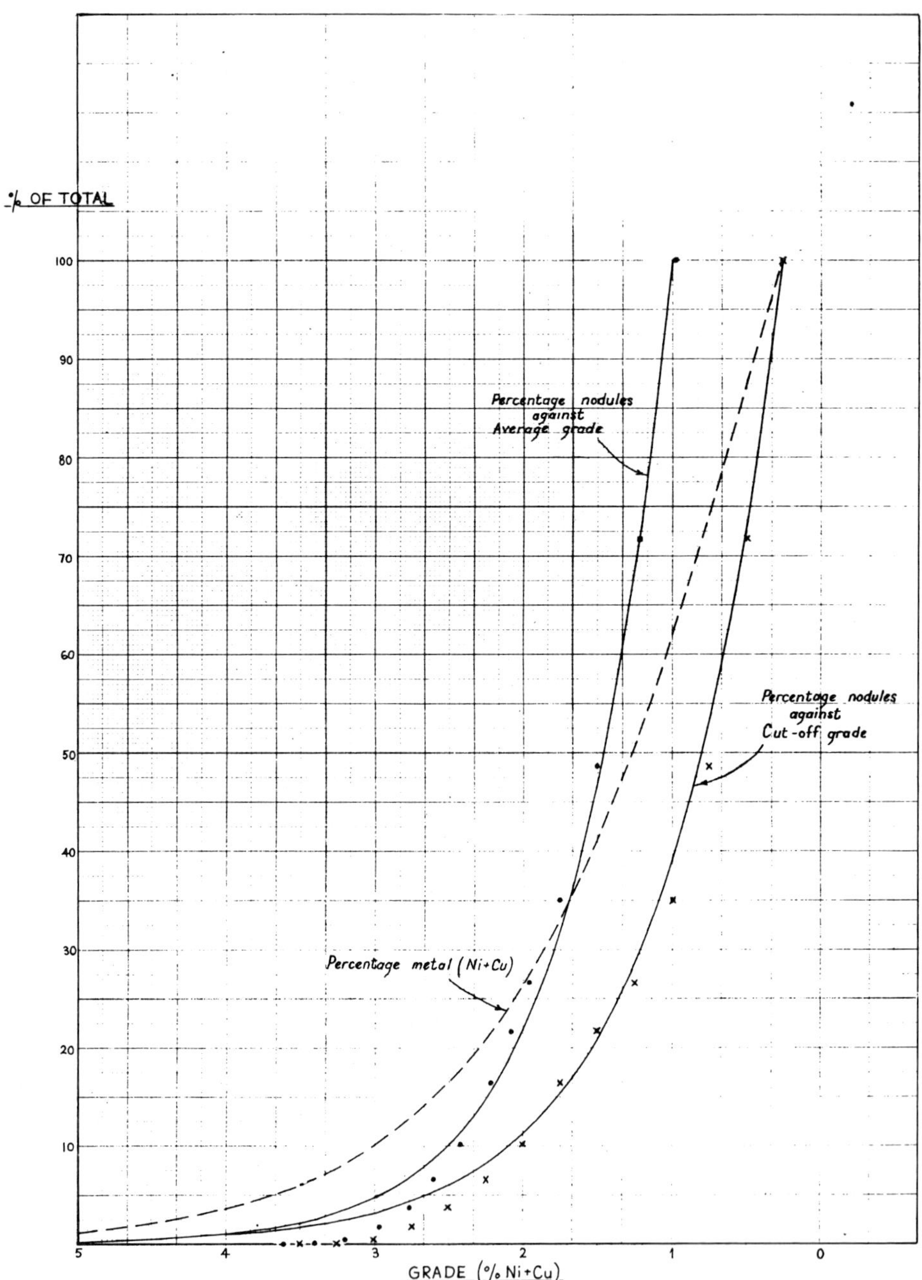

FIGURE 3. Cumulative Frequency By Grade In Each Ocean.

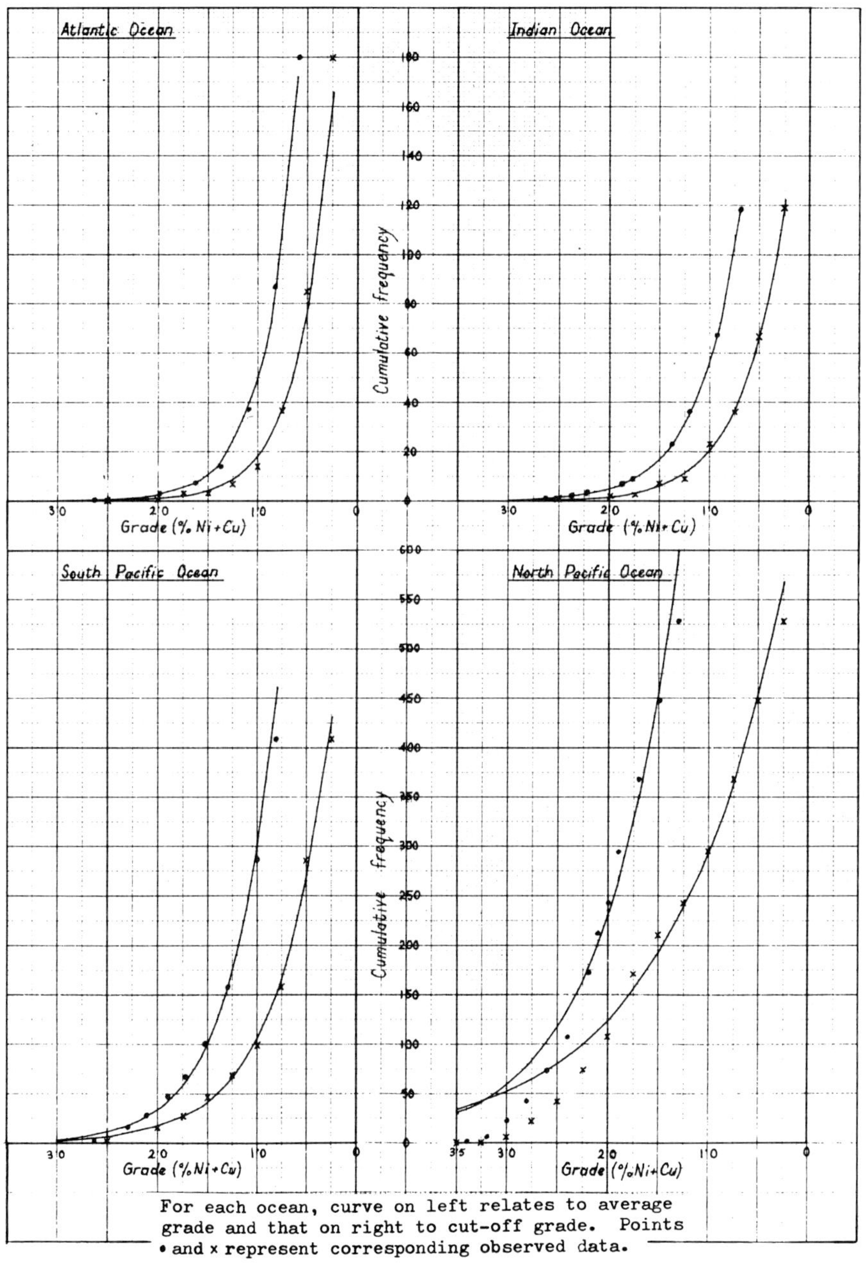

For each ocean, curve on left relates to average grade and that on right to cut-off grade. Points • and x represent corresponding observed data.

FIGURE 4. Frequency Distribution of Grade: Percentage Ni + Cu.

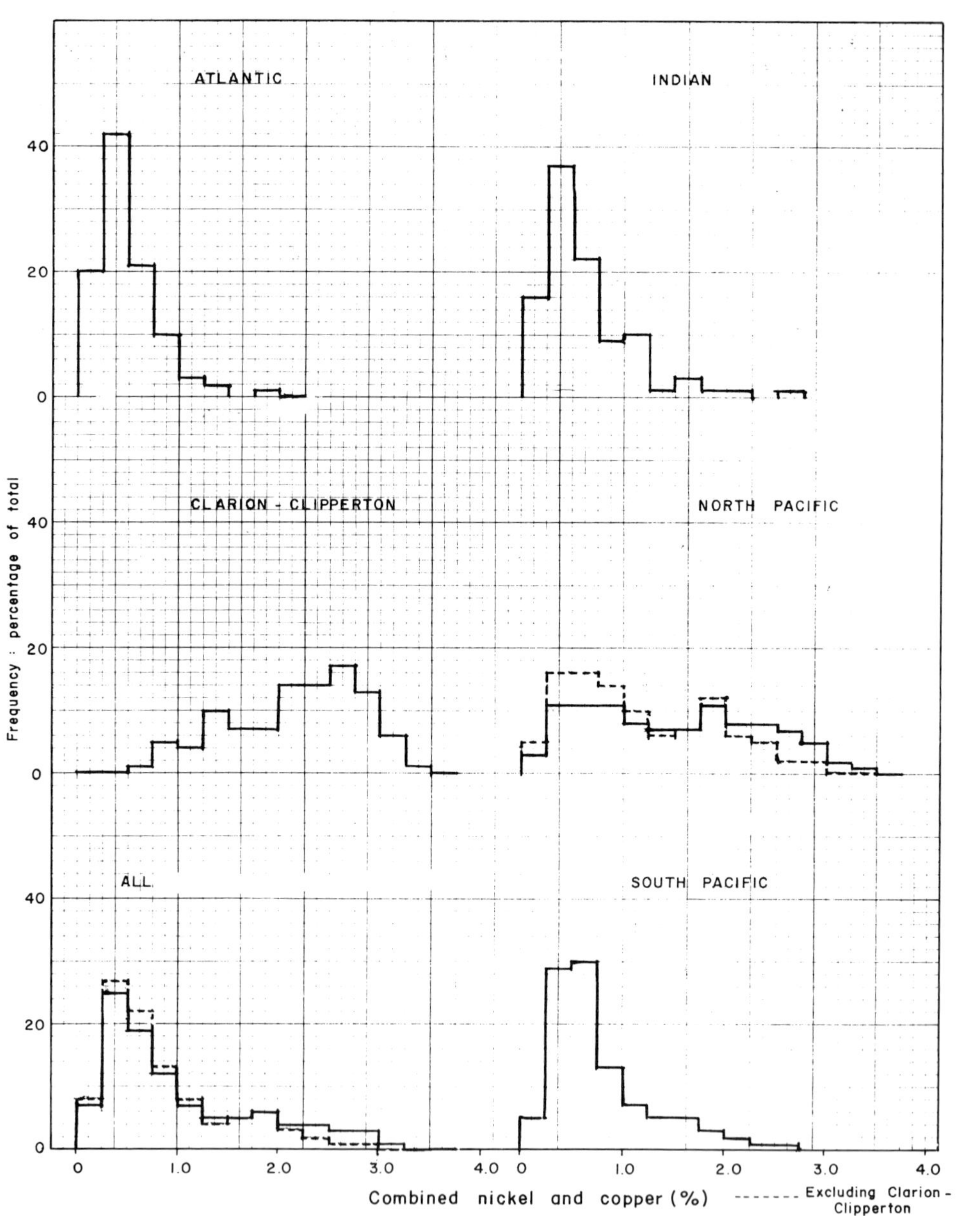

5. Nickel and Copper

The separate distributions of nickel and copper grades (Tables 8, 9 and Figures 6, 7) in individual and in all oceans are similar to those for the two metals combined. For both metals there is a good fit between the observed and expected (Table 7) values, as shown in Figure 5. As illustrated by the histograms (which both include and exclude all the Clarion-Clipperton data), the distribution that appears to be characteristic of the Clarion-Clipperton area again seems to be echoed in the bimodal distribution in the rest of the North Pacific and, more notably in the case of nickel, in the distribution for all oceans.

TABLE 6.
All Oceans: Comparison of Expected and Observed Frequencies at Each Grade

Grade	(a)		(b)		(c)	
(%Ni + Cu)	Expected	Observed	Expected	Observed	Expected	Observed
>3.50	14	0	57	1	21	0
3.25—<3.50	5	1	15	4	8	2
3.00—<3.25	8	1	20	15	10	4
2.75—<3.00	11	11	25	39	14	17
2.50—<2.75	16	16	33	53	19	23
2.25—<2.50	22	30	41	62	26	36
2.00—<2.25	31	38	52	68	36	44
1.75—<2.00	44	75	67	91	49	78
1.50—<1.75	61	61	86	76	67	64
1.25—<1.50	86	56	108	77	92	60
1.00—<1.25	122	102	131	111	125	104
0.75—<1.00	171	166	179	177	171	168
0.50—<0.75	241	286	225	288	234	287
0.25—<0.50	338	349	289	350	319	349
Totals	1170	1192	1335	1412	1191	1236

(a) Excluding Clarion-Clipperton. The expected frequencies are calculated (by differencing) from:

$$\log T = \frac{5.420 - G_c}{1.685}$$

(b) Including Clarion-Clipperton. The expected frequencies are calculated (by differencing) from:

$$\log T = \frac{7.647 - G_c}{2.367}$$

(c) Including Clarion-Clipperton (weighted data). The expected frequencies are calculated (by differencing) from:

$$\log T = \frac{5.920 - G_c}{1.843}$$

TABLE 7.
All Oceans*: Derived Equations Relating Cumulative Frequency (T) to Cutoff Grade (G_c) For Different Metals

Metal	Cutoff grade (% stated metal)
Nickel	G_c = 3.712 — 1.152 log T
Copper	G_c = 2.554 — 0.822 log T
Cobalt	G_c = 1.430 — 0.435 log T
Nickel Equivalent**	G_c = 4.097 — 1.256 log T

*Including Clarion-Clipperton (weighted data)
**Ni + Cu_3

6. Cobalt

The distribution of cobalt grades at the 1370 stations for which these were held in the Scripps Institution of Oceanography Data Bank in June, 1977 is shown in Table 10 and Figure 8. The distributions are generally similar to those of nickel and copper; for example, nodules with the lowest cobalt content do not frequently follow the general relationship. They are different, however, in that the frequency decreases much more rapidly as the grade increases; at only 14 of the stations is a cobalt content of more than 1% recorded and over 90% have less than 0.5% cobalt, while in the Clarion-Clipperton area only about 3% of the nodules have more than 0.4%, and the cobalt content is within the narrow range of 0.1 and 0.3% at 80% of the stations. These distributions are consistent, of course, with the fact that most nodules have less cobalt than nickel or copper and with the view that those with a relatively high cobalt content occur in a different environment[8 9].

The curve (Figure 5) derived from the equation relating cutoff grade to cumulative frequency (Table 7) shows that there is also a good fit between the observed and expected values for this metal.

TABLE 8.
Summary of Data in Scripps Data Bank (June 1977): Nickel*

Nickel Content %	Number of stations (percentage of totals in brackets)									
	Atlantic Ocean				South	North Pacific Ocean			All oceans	
	North	South	All	Indian Ocean	Pacific Ocean	Clarion-Clipperton	Rest	All	All	Except Clarion-Clipperton
0 —0.125	6 (6)	29 (19)	31 (14)	24 (17)	16 (4)	1 (0)	20 (4)	21 (3)	92 (6)	91 (7)
0.125—0.250	35 (37)	30 (23)	65 (29)	25 (18)	62 (14)	1 (0)	30 (6)	31 (4)	183 (12)	182 (14)
0.250—0.375	22 (23)	29 (22)	51 (23)	29 (21)	107 (25)	2 (1)	65 (13)	67 (9)	254 (17)	252 (19)
0.375—0.500	18 (19)	14 (11)	32 (14)	24 (17)	83 (19)	2 (1)	70 (14)	72 (10)	211 (14)	209 (16)
0.500—0.625	9 (9)	9 (7)	18 (8)	14 (10)	54 (13)	7 (3)	74 (15)	81 (11)	167 (11)	160 (12)
0.625—0.750	2 (2)	8 (6)	10 (4)	13 (9)	30 (7)	13 (6)	39 (8)	52 (7)	105 (7)	92 (7)
0.750—0.875	2 (2)	5 (4)	7 (3)	7 (5)	21 (5)	18 (8)	31 (6)	49 (7)	84 (6)	66 (5)
0.875—1.000	—	4 (3)	4 (2)	1 (1)	12 (3)	14 (6)	19 (4)	33 (5)	50 (3)	36 (3)
1.000—1.125	—	—	—	1 (1)	13 (3)	25 (11)	50 (10)	75 (10)	89 (6)	64 (5)
1.125—1.250	—	3 (2)	3 (1)	1 (1)	9 (2)	20 (9)	55 (11)	75 (10)	88 (6)	68 (5)
1.250—1.375	—	2 (2)	2 (2)	—	10 (2)	46 (21)	28 (6)	74 (10)	86 (6)	40 (3)
1.375—1.500	—	—	—	2 (1)	4 (1)	30 (14)	13 (3)	43 (6)	49 (3)	19 (1)
1.500—1.625	1 (1)	—	1 (0)	—	5 (1)	20 (9)	8 (2)	28 (4)	34 (2)	14 (1)
1.625—1.750	—	—	—	—	3 (1)	13 (6)	3 (1)	16 (2)	19 (1)	6 (0)
1.750—1.875	—	—	—	—	1 (0)	5 (2)	1 (0)	6 (1)	7 (0)	2 (0)
1.875—2.000	—	—	—	—	—	4 (2)	1 (0)	5 (1)	5 (0)	1 (0)
Totals	95	129	224	141	430	221	507	728	1523	1302

Note: not all percentages add up to 100% due to rounding.

TABLE 9.
Summary of Data in Scripps Data Bank (June 1977): Copper*

Copper Content %	Number of stations (percentage of totals in brackets)									
	Atlantic Ocean			Indian Ocean	South Pacific Ocean	North Pacific Ocean			All oceans	
	North	South	All			Clarion-Clipperton	Rest	All	All	Except Clarion-Clipperton
0 —0.125	57 (9)	78 (60)	135 (60)	48 (34)	83 (19)	3 (1)	100 (20)	103 (14)	369 (24)	366 (28)
0.125—0.250	30 (32)	37 (29)	67 (30)	55 (39)	174 (40)	—	75 (15)	75 (10)	371 (24)	371 (28)
0.250—0.375	5 (5)	11 (9)	16 (7)	21 (15)	85 (20)	7 (3)	66 (13)	73 (10)	195 (13)	188 (14)
0.375—0.500	2 (2)	1 (1)	3 (1)	6 (4)	41 (10)	21 (10)	56 (11)	77 (11)	127 (8)	106 (8)
0.500—0.625	1 (1)	1 (1)	2 (1)	4 (3)	22 (5)	17 (8)	43 (8)	60 (8)	88 (6)	71 (5)
0.625—0.750	—	—	—	1 (1)	13 (3)	15 (7)	66 (13)	81 (11)	95 (6)	80 (6)
0.750—0.875	—	1 (1)	1 (0)	3 (2)	5 (1)	22 (10)	30 (6)	52 (7)	61 (4)	39 (3)
0.875—1.000	—	—	—	1 (1)	3 (1)	20 (9)	18 (4)	38 (5)	42 (3)	22 (2)
1.000—1.125	—	—	—	2 (1)	3 (1)	38 (17)	21 (4)	59 (8)	64 (4)	26 (2)
1.125—1.250	—	—	—	—	—	28 (13)	17 (3)	45 (6)	45 (3)	17 (1)
1.250—1.375	—	—	—	—	1 (0)	25 (11)	6 (1)	31 (4)	32 (2)	7 (1)
1.375—1.500	—	—	—	—	—	12 (5)	2 (0)	14 (2)	14 (1)	2 (0)
1.500—1.625	—	—	—	—	—	11 (5)	5 (1)	16 (2)	16 (1)	5 (0)
1.625—1.750	—	—	—	—	—	2 (1)	—	2 (0)	2 (0)	—
1.750—1.875	—	—	—	—	—	—	1 (0)	1 (0)	1 (0)	1 (0)
1.875—2.000	—	—	—	—	—	—	1 (0)	1 (0)	1 (0)	1 (0)
Totals	95	129	224	141	430	221	507	728	1523	1302

Note: not all percentages add up to 100% due to rounding.

TABLE 10.
Summary of Data in Scripps Data Bank (June 1977): Cobalt*

Cobalt Content %	Number of stations (percentage of totals in brackets)									
	Atlantic Ocean			Indian Ocean	South Pacific Ocean	North Pacific Ocean			All oceans	
	North	South	All			Clarion-Clipperton	Rest	All	All	Except Clarion-Clipperton
0 —0.1	8 (9)	31 (24)	39 (18)	26 (19)	37 (9)	7 (4)	55 (13)	62 (10)	164 (12)	157 (13)
0.1—0.2	12 (14)	36 (28)	48 (22)	51 (37)	80 (20)	44 (25)	112 (26)	156 (25)	335 (24)	291 (24)
0.2—0.3	27 (31)	28 (22)	55 (26)	32 (23)	86 (21)	97 (54)	113 (26)	210 (34)	383 (28)	286 (24)
0.3—0.4	20 (23)	13 (10)	33 (15)	16 (12)	98 (24)	24 (13)	80 (18)	104 (17)	251 (18)	227 (19)
0.4—0.5	10 (12)	9 (7)	19 (9)	8 (6)	63 (16)	4 (2)	29 (7)	33 (5)	123 (9)	119 (10)
0.5—0.6	5 (6)	3 (2)	8 (4)	1 (1)	22 (5)	2 (1)	19 (4)	21 (3)	52 (4)	50 (4)
0.6—0.7	2 (2)	2 (2)	4 (2)	—	6 (1)	—	10 (2)	10 (2)	20 (1)	20 (2)
0.7—0.8	1 (1)	4 (3)	5 (2)	2 (1)	3 (1)	—	8 (2)	8 (1)	18 (1)	18 (2)
0.8—0.9	—	—	—	1 (1)	—	—	4 (1)	4 (1)	5 (0)	5 (0)
0.9—1.0	1 (1)	1 (1)	2 (1)	1 (1)	—	—	2 (0)	2 (0)	5 (0)	5 (0)
1.0—1.1	—	1 (1)	1 (0)	—	2 (0)	—	2 (0)	2 (0)	5 (0)	5 (0)
1.1—1.2	—	—	—	—	—	—	1 (0)	1 (0)	1 (0)	1 (0)
1.2—1.3	—	—	—	—	1 (0)	—	—	—	1 (0)	1 (0)
1.3—1.4	—	—	—	—	—	—	1 (0)	1 (0)	1 (0)	1 (0)
1.4—1.5	—	1 (1)	1 (0)	—	1 (0)	—	1 (0)	1 (0)	3 (0)	3 (0)
1.5—1.6	—	—	—	—	—	—	1 (0)	1 (0)	1 (0)	1 (0)
1.6—1.7	—	—	—	—	—	—	—	—	—	—
1.7—1.8	—	—	—	—	—	—	—	—	—	—
1.8—1.9	—	—	—	—	1 (0)	—	—	—	1 (0)	1 (0)
1.9—2.0	—	—	—	—	—	—	—	—	—	—
2.0—2.1	—	—	—	—	—	—	—	—	—	—
2.1—2.2	—	—	—	—	—	—	—	—	—	—
2.2—2.3	—	—	—	—	1 (0)	—	—	—	1 (0)	1 (0)
Totals	86	129	215	138	401	178	438	616	1370	1192

Note: not all percentages add up to 100% due to rounding.

FIGURE 5. All Oceans: Cumulative Frequency of Different Metals by Grade.

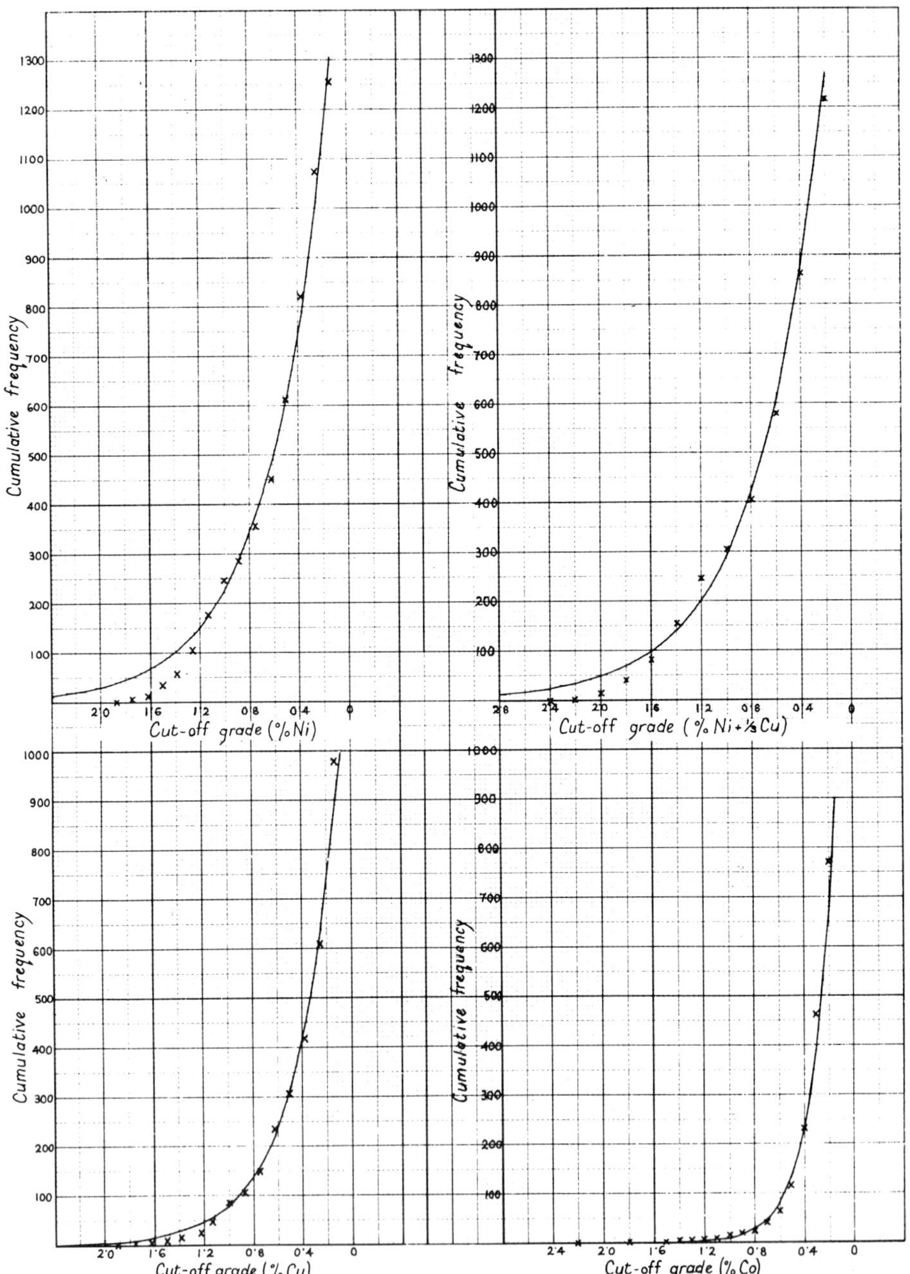

7. Nickel Equivalent

If the exploitation of manganese nodule deposits is to be based on the production of co-product nickel and copper, account should be taken of the different market values of these metals. If, for example, the value of nickel is three times that of copper then, following the same procedure, the distribution of 'nickel equivalent' grades calculated as Ni + $\frac{Cu}{3}$ (Table 11) is of more practical relevance than the distribution of the unweighted combined nickel and copper content. Figure 5 shows that the curve fitted to the corresponding equation (Table 7) provides a similar fit to the observed data as that provided by the curves showing the cumulative percentages of unweighted combined nickel and copper grade (Figure 2).

TABLE 11.
Summary of Data in Scripps Data Bank (June 1977): Nickel + 1/3 Copper*

	Number of stations (percentage of totals in brackets)		
Ni + Cu/3 content %	Clarion-Clipperton zone	All Oceans except Clarion-Clipperton	All Oceans including Clarion-Clipperton (weighted)
0.2		354 (30)	354 (29)
0.2 — 0.4	4 (2)	280 (24)	281 (23)
0.4 — 0.6	11 (5)	173 (15)	175 (14)
0.6 — 0.8	21 (10)	97 (2)	101 (8)
0.8 — 1.0	16 (7)	57 (5)	60 (5)
1.0 — 1.2	25 (11)	85 (7)	90 (7)
1.2 — 1.4	35 (16)	65 (6)	72 (6)
1.4 — 1.6	42 (19)	35 (3)	43 (4)
1.6 — 1.8	39 (18)	18 (2)	26 (2)
1.8 — 2.0	17 (8)	7 (1)	10 (1)
2.0 — 2.2	7 (3)	1 (0)	2 (0)
2.2 — 2.4	1 (0)	0	0
2.4 — 2.6	—	—	—
2.6 — 2.8	—	—	—
Totals	218	1172	1214

*Not all percentages add up to 100% due to rounding.

8. Conclusions

The examination of the frequency distribution of grades (expressed as combined nickel and copper) based on a substantially larger number of samples tends to confirm the tentative conclusion reached by Healing and Archer in 1975 that the distribution is similar to the grade-tonnage relationship suggested by Lasky. Furthermore, this seems to apply when data from the Clarion-Clipperton area, widely regarded as being the largest area with high grade nodules, are included. A similar relationship applies to cobalt and to grade expressed as the nickel equivalent, as well as to nickel and copper separately.

While possibly advancing the hypothesis, it is premature to suggest that these results demonstrate the existence of any underlying natural law or even prove the existence of a Lasky grade-tonnage relationship. However, it can be said that the usefulness of the equations (or of the curves generated by them) as practical tools for quantitative resource assessments, is enhanced. Generalized conclusions can be drawn from them with slightly more confidence. The degree of confidence is related to the number of samples available from an area. Thus, conclusions about the distribution of grade in the Atlantic and Indian Oceans must be even more tentative than those for the North and South Pacific, and even the relationships between grade and frequency in all the oceans taken together must still be used with care.

The equations for combined nickel and copper in all oceans including weighted data for Clarion-Clipperton (Table 4) can be used to indicate the combined effect of increases in the number of nodules (which provides a measure of tonnage provided that grade and abundance vary independently) and decreases in their average or cutoff grade, on the quantity of metal they contain. Thus of the total samples (except those below a cutoff of 0.25%), about 15% occur with a cutoff of 1.76% while 45% occur with a cutoff of 0.88%. For the same cutoffs, the proportion of contained metal increases from about 33% to 68%. That is to say that, in this case, halving the cutoff grade trebles the number of samples (or the tonnage) but only doubles the amount of metal.

9. Limitations

This contribution attempts only to indicate the ways in which nickel, copper and cobalt are distributed in manganese nodule deposits. Inevitably it depends on several assumptions. In many cases the Lasky distribution provides a good fit; other equations which might provide an even better fit have not been tested. No attempt has been made to fit an equation to the distribution of nickel and copper, either separately or combined, in the Clarion-Clipperton area. In particular, explanations have not been offered for the different distributions that are illustrated; this is in the domain of specialized geochemists and mineralogists. This is not to say that explanations are not available, for example, for the bimodal distribution of nickel and copper in the Clarion-Clipperton area[10], while the higher nickel and copper values that the equations suggest should be present have been found in thin layers of nodules[11].

APPENDIX A

WEIGHING OF CLARION-CLIPPERTON DATA

As defined, the Clarion-Clipperton area is about 3.4 million km^2. The data tabulated by Frazer[12] suggest that nodules may be present in about half of this area, that is about 1.7 million km^2. This is about 3% of the total area of the seabed occupied by nodules, if Archer's estimate of 54 million km^2 is accepted. On the other hand, an earlier estimate[13], based on data published by Schultze-Westrum[14], indicated that about 90% of the western half of the Clarion-Clipperton area may be occupied by nodules. If this applies to the whole of the Clarion-Clipperton area, about 3 million km^2 would be occupied by nodules. This area would account for about 5.6% of the world total. The appropriate weighting (page 47) would then be by taking about one-third of the samples from this area. The equation for cutoff grade for all oceans (Table 4) recalculated on the latter basis would be:

$$G_C = 6.225 - 1.938 \log T$$

Among the effects would be that the cumulative frequencies above cutoffs of 1.76% and 0.88% combined nickel and copper would increase from 15% and 45% to 17% and 47%, respectively. This effect is small, compared with other sources of error.

APPENDIX B

GOODNESS OF FIT

In the text of the present paper (and in Healing and Archer) the fit of the Lasky curves to the observed data is generally described as "good". This is a subjective assertion based only upon visual inspection of the relevant tabulations and graphs.

Objective, quantitative tests of goodness of fit were not attempted because (a) at this early, explorative stage, visual impressions were considered to be enough and freely open to the judgement of readers as well as the authors, and (b) time has not allowed the respective validities of the various test options to be thoroughly investigated.

Nevertheless, it is recognized that some objective, qualitative test, however superficial, might be of immediate assistance for comparative purposes. For all the principal relationship in Healing and Archer and the present paper a chi-squared (X^2) routine has therefore been carried out on the differences between the expected (Lasky) and observed cumulative frequencies. That is $(E-O)^2/E$, where E = the expected and O = the observed cumulative frequencies, has been summed over the eight values (in each case) on which the Lasky curves were based. The value of chi-squared at the 5% probability level, with seven degrees of freedom is 14.07. For the purpose of this exercise it will thus be taken that where the differences between the expected and observed cumulative frequencies yield a chi-squared value of less than 14.07, such differences may be attributed to chance only and are not large enough to invalidate the hypothesis of a Lasky relationship.

The first results are as follows:

	Chi-squared		Chi-squared
Atlantic Ocean (Ni + Cu)		All Oceans (a)	
1975 data	6.25	(Ni + Cu)	
1977 data	3.72	1975 data	0.75
Indian Ocean (Ni + Cu)		1977 data	10.72
1975 data	0.39	All Oceans (b)	
1977 data	1.11	1977 data	
S. Pacific Ocean (Ni + Cu)		Ni + Cu	11.63
1975 data	0.96	Ni	10.94
1977 data	3.79	Cu	10.83

N. Pacific Ocean (Ni + Cu) (a)		Ni + C	79.21
1975 data	1.97	Co	36.10
1977 data	13.30		
N. Pacific Ocean (Ni + Cu) (b)			
1977 data	9.02		

(a) Excluding Clarion-Clipperton data
(b) Including Clarion-Clipperton data

Preliminary conclusions are that the grade distributions for combined nickel and copper, in each ocean and for all oceans, all pass the superficial chi-squared test applied to them. In each case the differences between the expected and observed cumulative data are small enough (only just small enough in certain instances) to be attributable to chance, i.e. not large enough to reject the Lasky hypothesis. This result also applies to the all-oceans distributions of nickel and copper grades separately, but not to the distribution of "nickel equivalent" ($Ni + Cu_3$) or to the cobalt grades.

For combined nickel and copper grades the goodness of fit found in 1975 by Healing and Archer compares favourably (i.e. smaller differences between expected and observed cumulative frequencies) than in 1977 (the present paper) for each ocean other than the Atlantic and for all oceans together (excluding Clarion-Clipperton). The goodness of fit found in 1977 (this paper) was also slightly better with Clarion-Clipperton excluded altogether than with Clarion-Clipperton included on a weighted basis. In all these cases, however, the goodness of fit was still enough not to invalidate the hypothesis of a Lasky distribution. Much further work needs to be done on testing goodness of fit, however, and these preliminary conclusions should be treated with caution.

APPENDIX C
PROPORTIONAL LASKY CURVES
(percentage of total nodules* above each cutoff grade)

For simplicity in the body of this paper most of the fitted Lasky curves have been presented in absolute terms, i.e. showing the number of stations reporting nodules at or above each grade. Exceptions are Figures 1 and 2, in which the curves are presented in relative terms, i.e. showing the proportions of total nodules* above each cutoff grade. The latter is a useful form of presentation as, *inter alia,* it enables comparisons to be made directly between any curves independently of the number of observations on which they are based, e.g. between the curves for individual oceans and for all oceans, and between the curves derived in 1975 and in this paper.

The transformation is a simple one. Given the Lasky relationship $G = G_c = K_3 - K_4 \log T$ (see Method above) then P, the percentage of total nodules* above each cutoff grade G_c is given by: $\log P = 2 + (0.25 - G_c)/K_4$.

Equations in this form for all the grade (% Ni + Cu) relationships derived by Healing and Archer in 1975 and in 1977 (this paper) have been calculated as follows:

*Ignoring all nodules containing less than 0.25% Ni + C

	K_4	Log P
Atlantic Ocean		
1975 data	0.995	2.251 — 1.005 G_c
1977 data	0.775	2.323 — 1.290 G_c
Indian Ocean		
1975 data	1.251	2.200 — 0.799 G_c
1977 data	0.961	2.260 — 1.041 G_c
S. Pacific Ocean		
1975 data	1.789	2.140 — 0.559 G_c
1977 data	1.230	2.203 — 0.813 G_c
N. Pacific Ocean (a)		
1975 data	1.962	2.127 — 0.510 G_c
1977 data	2.334	2.107 — 0.429 G_c
N. Pacific Ocean (b)		
1977 data		
N. Pacific Ocean (c)		
1977 data	2.657	2.094 — 0.376 G_c
All Oceans (a)		
1975 data	1.721	2.145 — 0.581 G_c
1977 data	1.685	2.148 — 0.593 G_c
All Oceans (b)		
1977 data	2.367	2.106 — 0.422 G_c
All Oceans (c)		
1977 data	1.843	2.136 — 0.543 G_c

(a) Excluding Clarion-Clipperton data.
(b) Including Clarion-Clipperton data.
(c) Including Clarion-Clipperton data (weighted).

Notes and References

[1] Healing, R.A. and A.A. Archer, 'Manganese Nodules — An Example of Lasky Distribution?', *Resources Policy,* 1975, Vol. 1, No. 6, p. 306-312.

[2] Lasky, S.G., 'How Tonnage and Grade Relations Help Predict Ore Reserves', *Engineering and Mining Journal,* Vol. 151, No. 4, 1950, p. 81-85.

[3] Frazer, J.Z. and G. Arrhenius, 'World-wide Distribution of Ferromanganese Nodules and Element Concentrations in Selected Pacific Ocean Nodules', *International Decade of Ocean Exploration,* Technical Report No. 2, 1972, 51 p. (Unpublished manuscript).

[4] Horn, D.R., M.N. Delach and B.M. Horn, 'Metal Content of Ferromanganese Deposits of the Oceans', *International Decade of Ocean Exploration,* Technical Report No. 3, 1973, 51 p. (Unpublished manuscript).

[5] Monget, J.M., J.W. Murray and J. Mascle, 'A World-wide Compilation of Published Multicomponent Analyses of Ferromanganese Concretions', *International Decade of Ocean Exploration,* Technical Report No. 12 (NFS Grant 33616), 1976, 127 p.

[6] Archer, A.A., Chapter III.2. this publication.

[7] Menard, H.W. and J.Z. Frazer, 'Inverse Correlation between Grade and Abundance of Manganese Nodules', (In press).

[8] Cronan, D.S. and J.S. Tooms, 'The Geochemistry of Manganese Nodules and Associated Pelagic Deposits from the Pacific and Indian Oceans', *Deep-Sea Resources,* Vol. 16, 1969, p. 335-359.

[9] Howarth, R.J., D.S. Cronan and G.P. Glasby, 'Non-linear Mapping of Regional Geochemical Variability of Manganese Nodules in the Pacific Ocean', *Bulletin and Transactions Institution of Mining and Metallurgy,* Section B, Vol. 86, 1977, p. B4-B8.

[10] Friedrich, G. and W. Pluger, 'The Distribution of Manganese, Iron, Cobalt, Nickel, Copper and Zinc in Manganese Nodules from Different Fields', *Meerestechnik,* Vol. 5, No. 6, 1974, p. 203-206. (In German; English abstract).

[11] Margolis, S.V. and R.G. Burns, 'Pacific Deep-Sea Manganese Nodules: Their Distribution, Composition and Origin', *Annual Review Earth and Planetary Science Letters,* 1976, Vol. 4, p. 229-263.

[12] Frazer, J.Z., 'Manganese Nodules Reserves: An Updated Estimate', *Mar., Min.,* Vol. 1, Nos. 1/2, November 1977, p. 103-123.

[13] Archer, A.A., Prospects for the Exploitation of Manganese Nodules: The Main Technical, Economic and Legal Problems, Paper presented at the I.D.O.E. Workshop, Suva, Fiji, September 1975, *Technical Bulletin No. 2, CCOP/SOPAC,* 1976, p. 21-38.

[14] Schultze-Westrum, H.H., 'The Station and Cruise Pattern of the R/V Valdivia in Relation to the Variability of Manganese Nodule Occurrences', Morgenstein, M. (Editor), *Papers on the Origin and Distribution of Manganese Nodules in the Pacific and Prospects for Exploration,* International Symposium, Honolulu, Hawaii, 23-25 July 1973, p. 145-149.

CHAPTER V

STATUS OF THE MARINE GEOLOGICAL ACTIVITIES OF THE GEOLOGICAL SURVEY OF JAPAN

Three years have passed since the establishment of the Marine Mineralogy Department and the launching of the research vessel Hakurei-Maru (1821 gross tons) in 1974. The Marine Mineralogy Department consists of three sections: Marine Geology, Marine Mineral Resources, and Marine Geophysics. Each section has been actively conducting various research projects with considerable results. The Department has grown as one of the leading marine geological research centres in Japan.

Recent research activities include two main projects using the research vessel Hakurei-Maru, "Marine geological investigations on continental shelves and shores around Japan" and "Investigations on deep-sea manganese nodule deposits", and two other projects in nearshore areas using chartered smaller vessels, "Environmental studies on bottom sediments" and "Investigations on submarine sand and gravel resources in shallow water". The survey methods applied in the Hakurei-Maru cruises and the results of her cruises during 1976 and 1977 are shown in Table 1 and Table 2 respectively. In addition, the Department has been contributing to a wide range of international cooperation in the field of marine geology in various ways.

Opportunities for training on-board the research vessel Hakurei-Maru during some parts of her cruises around Japan have been provided each year for at least ten foreign participants in the group training course in offshore prospecting sponsored by the Japanese Government. Two individual foreign trainees have been accepted in the Hakurei-Maru cruises both for an area around Japan and for the Central Pacific Basin for manganese nodule studies, respectively. A hydrologist from Western Samoa joined the GH 77-1 cruise from Apia to Funabashi, organized in January-March 1977. A scientist from Canada also participated in a Hakurei-Maru cruise to the Pacific area off Northeast Japan to study sedimentology in 1976.

As one of the international cooperative programmes, some geologists of the Marine Department will join in the Glomar Challenger IPOD Drilling cruises in the Western Pacific areas from the fall of 1977 to the beginning of 1978. Also, the scientists of the Marine Geology Department have attended a number of meetings under bilateral or multilateral governmental programmes, or of international scientific associations.

1. Marine Geological Investigations on Continental Shelves and Slopes Around Japan (First five year project, 1974-1978)

This project aims at mapping both subbottom geology and surface sediments of the areas around Japan Islands, resulting in publications of reconaissance submarine geologic maps on a scale of 1:1,000,000 for extensive areas, and also of detailed geologic and surface sediment maps on a scale of 1:200,000 in selected offshore areas (Figure 1).

Reconnaissance studies have been concentrated in the areas of the Pacific side during 1974-1976, and then expanded to the areas of the Sea of Japan and the Okhotsk during

TABLE 1
Method of Marine Geology Investigation Aboard R/V Hakurei-Maru.

Methods / Projects	Investigation around the Japanese Islands	Manganese investigation
Positioning	NNSS, Decca, Loran C/A	NNSS, Loran C
Geophysical methods		
Gravity measurement	On-board gravity meter	On-board gravity meter
Magnetic measurement	Proton magnetometer	Proton magnetometer
Depth sounding/topography	Precision depth recorder (12 kHz PDR)	Precision depth recorder (12 kHz PDR)
Continuous seismic reflection profiling	Air gun, sparker	Air gun
Seismic refraction	Sono-radio buoy	Sono-radio buoy
Sediment profiling	3.5 kHz PDR (subbottom prefiler)	3.6 kHz PDR (subbottom profiler)
Side looking survey	Side scan sonar	
Geological methods		
Sediment sampling	Smith-McIntyre grab, piston center, and gravity corer	Okean 70 grab, free fall grab, piston corer, and free fall corer
Rock sampling	Submersible rock drill, dredge, and rock corer	Dredge, rock corer
Sea bottom observation	Deep sea camera and diving	Deep sea camera, free fall camera, and deep sea television
Others		
Water sampling and measurement	Water sampler (Nansen bottle) STD	Water sampler (Nansen bottle) STD
On-board instrumental analysis	X-ray diffractometer, X-ray fluorescence spectrometer, radiography instrument, and others	

TABLE 2.
Hakurei-Maru Cruises by GSJ in 1976 and 1977.

GH 76-1 Cruise	January 10—March 9, 1976 (A. Mizuno) Central Pacific Basin, for manganese nodule deposits research.
GH 76-2 Cruise	April 17—June 4, 1976 (E. Honza) Off northern Honshu and Hokkaido (Pacific shelf, slope and trench), for geological mapping (1:1,000,000).
GH 76-3-1 Cruise	June 15—June 30, 1976 (E. Honza) Off Hachinohe (Pacific coast of northern Honshu), for geological mapping (1:200,000).
GH 76-3-2 Cruise	July 1—July 21, 1976 (E. Inoue) *Ibid.*, for sedimentological mapping (1:200,000).
GH 76-3-3 Cruise	July 22—July 30, 1976 (E. Inoue) Nishi-tsugaru Basin in Japan Sea coast of northern Honshu, for geological mapping (1:200,000).
GH 76-3-4 Cruise	July 31—August 4, 1976 (J. Chujo) Off Kinkazan (Pacific shelf, slope and trench), for a part of geological mapping (1:1,000,000).
GH 77-1 Cruise	January 12—March 12, 1977 (T. Moritani) Central Pacific Basin, for manganese nodule deposits research.
GH 77-2 Cruise	April 19—May 28, 1977 (E. Honza) The Sea of Japan off southwest Japan and around Oki Islands, for geological mapping (1:1,000,000 and 1:200,000).
GH 77-3-1 Cruise	June 14—July 9, 1977 (E. Honza) The Sea of Okhotsk and the Sea of Japan off Hokkaido, for geological mapping (1:1,000,000)
GH 77-3-2 Cruise	July 10—August 6, 1977 (E. Inoue) Off Nishi-tsugaru in the Sea of Japan, for sedimentological mapping (1:200,000).
GH 77-3-3 Cruise	August 6—August 12, 1977 (J. Chujo) The Sea of Japan to Pacific, for technical training of foreign students and supplementary research of geological mapping.

1977 (Figure 2). Through these works, we have obtained a number of geological and geophysical data over the whole area around Japan except the central part of the Sea of Japan, which is left for next year's cruises.

Detailed studies have been done so far with considerable collection of the geological and geophysical data for some shelf and slope areas, such as off western Kyushu, off Kii Peninsula, off South Kwanto, off Hachinohe, off Nishitsugaru, and around the Oki Islands.

To improve the bottom sampling technique, a submersible type rock drill machine capable of obtaining a maximum 75cm long core of the bed rock has been used in some stations for the detailed mapping survey on a scale of 1:200,000, in addition to the conventional methods of dredge and rock corer.

In 1978, the final year of the present five year project, the Hakurei-Maru cruises during April-August will be carried out in the central part of the Sea of Japan, at the end of which the whole of the continental margin around Japan will be completely covered.

FIGURE 1. Five-year Program (1974-1978) of the Geological and Sedimentological Research Around the Japanese Islands.

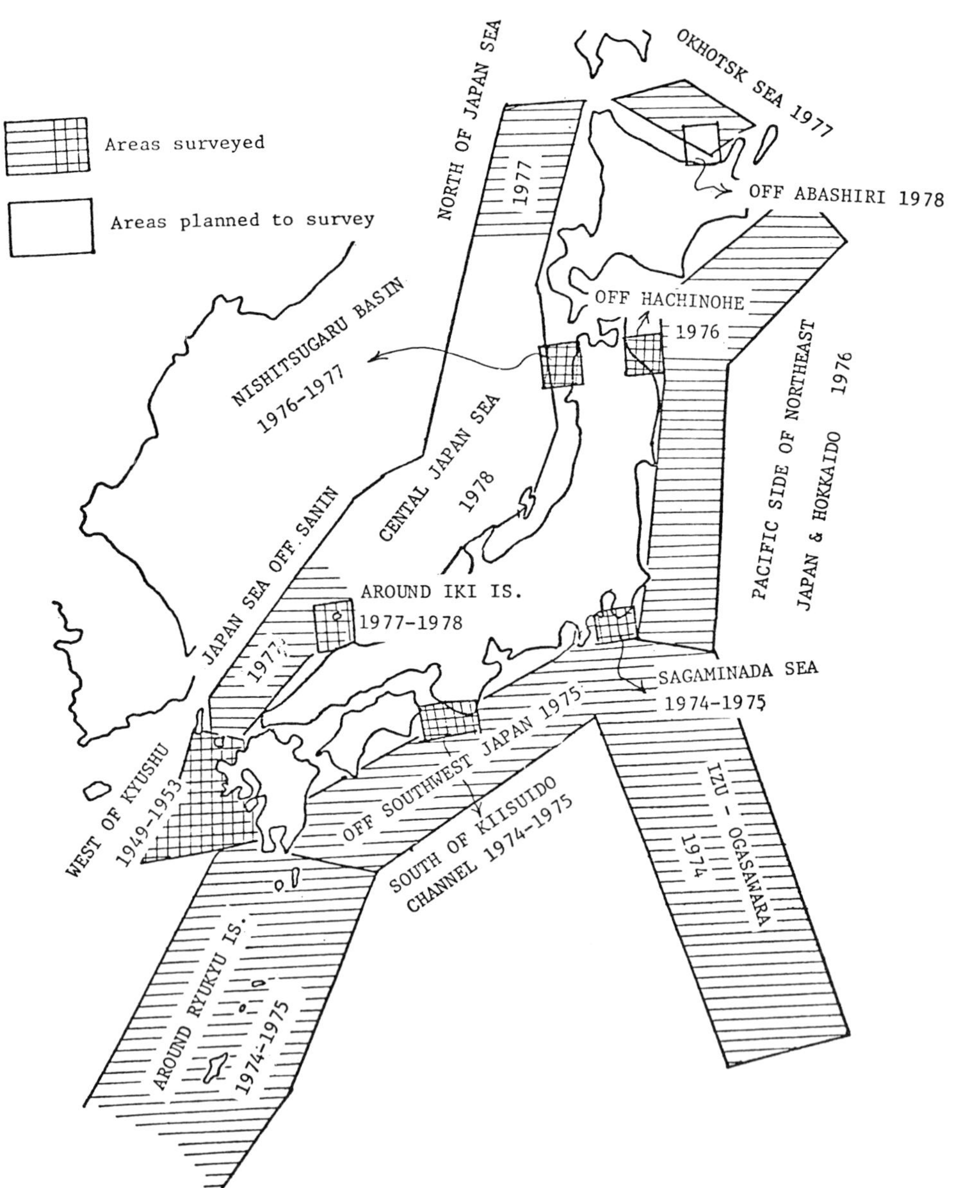

FIGURE 2. Surveyed Areas and Geophysical Track Lines Around the Japanese Islands, 1974-1977.

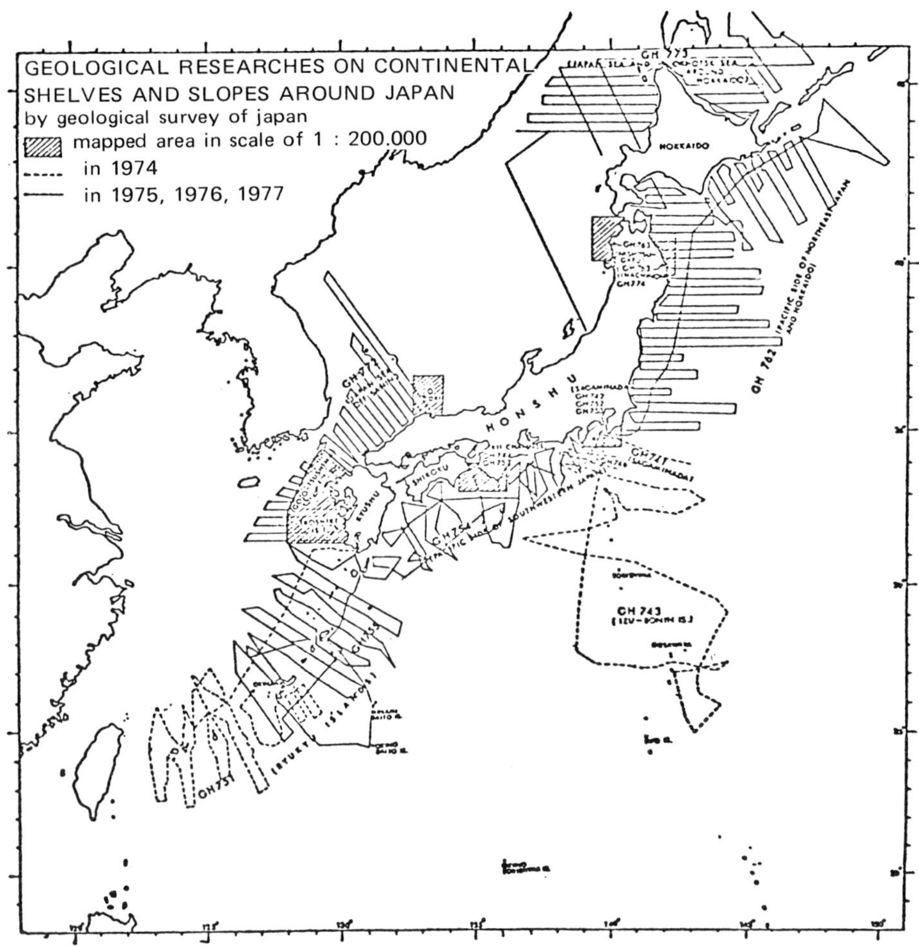

2. Investigations on Deep-Sea Manganese Nodule Deposits (Five year project, 1974-1978)

This project aims at providing fundamental scientific data on manganese nodule genesis in relation to topography, sedimentology and other geologic factors as basic information for their exploration and exploitation.

The main area has been the Central Pacific Basin since 1974. Three cruises of the Hakurei-Maru during 1974-1977 covered an extensive area of approximately 1,500km x 500km, which extends from a siliceous ooze zone in the south to a red clay zone in the north.

Through these cruises various geological, sedimentological and geophysical data of the Central Pacific Basin have been collected, and also both abundance of manganese nodules and their mineralogical and chemical compositions have been better delineated, particularly in relation to nodule morphology, submarine topography, and acoustic stratigraphy. These results, however, show the overall tendencies interpreted from the field data on the widely spaced survey lines and stations. Therefore, more detailed surveys with densely spaced samplings and geophysical measurements are necessary in further studies. Some preliminary research results were published[1] and the detailed data were given in regular publications, Cruise Report, No. 4 and No. 8 (Table 3), and also in other individual papers concerned.

In 1978, the GH 78-1 cruise is scheduled in the western part of the Central Pacific Basin during January-March 1978. Also, in 1979, the final year of the present project, another GH cruise will be planned to cover the whole area.

TABLE 3.
Main Publications Concerning Marine Geology and Marine Mineral Resources

Cruise Report	
No. 1	Deep Sea Mineral Resources Investigations in Northwest Pacific, November-December 1972 (1974)
No. 2	Goto-nada Sea and Tsushima Strait Investigations, Northwestern Kyushu, 1972-1973 (1975)
No. 3	Sagami-nada Sea Investigations, April-May 1974, GH 74-1 and -2 Cruises (1975)
No. 4	Deep Sea Mineral Resources Investigation in the Eastern Central Pacific Basin, August-October 1974, GH 74-5 Cruise (1975)
No. 5	Izu-Ogasawara (Bonin) Arc and Trench Investigations, June and October-November 1974, GH 74-3 and -6 Cruises (1976)
No. 6	Ryukyu Island (Nansei-Shoto) Arc, GH 75-1 and 75-5 Cruises, January-February and July-August 1975 (1976)
No. 7	Geological Investigation of Japan and Southern Kurile Trench and Slope Areas, GH 76-2 Cruise, April-June 1976 (1977)
No. 8	Deep Sea Mineral Resources Investigation in the Central-eastern part of Central Pacific Basin, January-March 1976, GH 76-1 Cruise. (in press)
No. 9	Geological Investigations off Pacific Coast of Southwestern Japan, GH 75-4 Cruise, June 1975. (in press)
Marine Geology Map Series	
No. 1	Submarine Geological Map around Koshikijima Islands 1:200,000 (1975)
No. 2	Bottom Sediment Map, Tsushima & Goto Islands Area 1:200,000 (1975)
No. 3	Submarine Geological Map of Sagami-nada Sea and Its Vicinity 1:200,000 (1976)
No. 4	Sedimentological Map of Sagami-nada Sea and Its Vicinity 1:200,000 (1976)
No. 5	Marine Geological Map of the South of Kii Strait 1:200,000 (1976)
No. 6	Sedimentological Map of the South of Kii Strait 1:200,000 (1976)
No. 7	Geological Map around Ryukyu Arc 1:1,000,000 (1977)
No. 8	Geological Map Off Outer Zone of Southwest Japan 1:1,000,000 (1977)

FIGURE 3. Research Areas for Manganese Nodule Deposits by the Geological Survey of Japan 1970-1974 and 1975-1978.

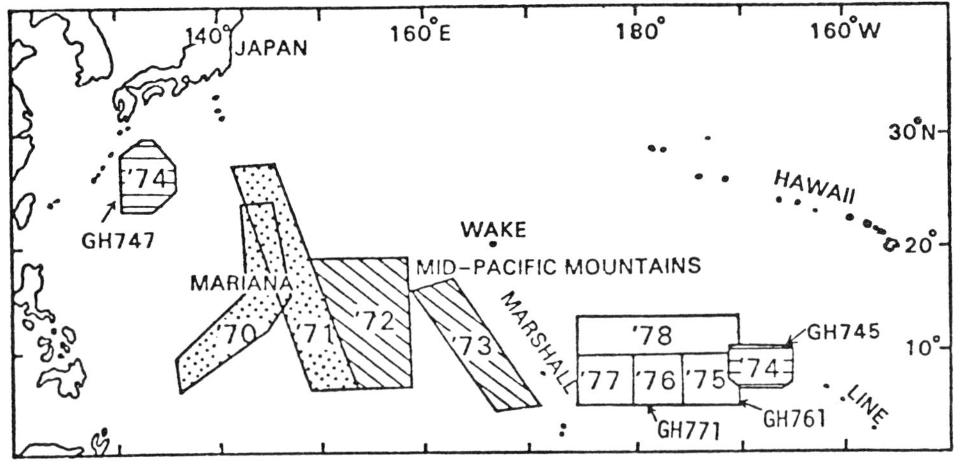

3. Environmental Studies on Bottom Sediments

The first three-year project entitled "Fundamental studies on the survey techniques for the polluted bottom sediments" has been carried out from 1974 to 1976, for the Seto Inland Sea areas. The developed survey techniques include those of the survey of the sediment distribution by acoustic equipment, the sampling of undisturbed sediments, the determination of sedimentation rate during the past 100 years by ^{210}Pb dating method, etc.

On the basis of the results of the former project, a new three-year project, "Studies on the sedimentation mechanism of the polluted bottom sediments", has been initiated in 1977 for the purpose of clarifying the origin of sediments and history of the sedimentary environments, by studying distribution of sediments relating to changes in their thickness, minor changes in seabottom topography, suspended materials, and physical and chemical properties of the sediments.

Three types of representative model fields of bay areas have been selected for this project from subarctic, temperate and subtropical zones, respectively, along Japan Islands. Since 1977, two areas of Shibushi Bay and Suruga Bay, both in the temperate zone, are being studied.

4. Investigations on Submarine Sand and Gravel Resources

This smaller scale project has been conducted since 1975, to provide basic data with respect to prospecting of sand and gravel in nearshore shallow waters at the depth of tens-of-metres, studying composition and physical properties of the sediments, submarine topography and quaternary history.

The areas already surveyed are off the northern coast of Kyushu (1975), off the western coast of Kyushu (1976), and off the western coast of Okinawa (1977).

The on-board survey methods of these projects including those of the above-mentioned environmental studies, consist mainly of depth sounding, bottom observation by means of a side scan sonar, subbottom profiling by means of a boomer, and bottom sampling by a Smith-McIntyre grab.

5. Publication Programme

The results of the above-stated projects and related studies of the Geological Survey of Japan are published in a series of Cruise Reports and a Marine Geology Map, and also as various individual papers in journals.

The Cruise Report, published at the end of each cruise, is edited by the chief scientist (and cochief scientist) and comprises many papers contributed by the on-board and non-onboard scientists concerned. Up to now, seven issues have been published and two are now in print (Table 3).

The Marine Geology Map series has two types of maps. One is a Marine Geological Map, each sheet issue of which consists usually of a geologic map, at magnetic anomaly map, and a gravity anomaly map. The other is the Sedimentological Map, which contains a granulometric map, and a coarse fraction composition map, etc. Eight sheet issues of the Marine Geology Map Series have been published so far (Table 3).

Notes and References

[1] A. Mizuno and T. Moritani; *Proceedings of Joint Meeting of the Mining and Metallurgical Institute of Japan and the American Institute of Mining Metallurgical and Petroleum Engineers,* Denver 1976.

PART III

THE MINE SITE

CHAPTER VI

PROBLEM ADDRESSED AND A SUMMARY OF THE DISCUSSION

1. Problem Addressed

Widely spaced reconnaissance surveys, of the kind described by Bastien-Thiry in Chapter I of The Basic Data lead to the identification of areas in which the composition and abundance of the nodules encourage more detailed prospecting. Such detailed prospecting is needed to identify areas in which the required grade and abundance are present in a topographically amenable area that is large enough to sustain a commercially viable mining operation. Such an area is called a mine-site.

Archer introduced his paper on "Resources and Potential Reserves of Nickel and Copper in Manganese Nodules" describing it as a revised version of an earlier attempt to estimate their extent. Although more than two years had elapsed, it was stressed that the data available remained inadequate, both in quality and reliability. Furthermore, the methods used involved assumptions about the randomness of the sampling, the distribution of the errors in the results obtained at the sample points and the relationship between grade and abundance. Each of these assumptions might be unjustified. Archer also emphasized that, as a result of the uncertainties involved, the estimates that were derived must be regarded as very tentative. They might, however, be of the right orders of magnitude. He suggested that it seemed likely to prove impossible for the Group of Experts to provide a single set of figures for use in any United Nations study of the development of the world economy. Archer also pointed out that although the estimates of potential reserves could be converted to estimates of the number of mine-sites by making further assumptions, this was not strictly an element in the Group's terms of reference.

Pasho, in his paper entitled "Determining Deep-Sea-Bed Mine-Site Area Requirements", presented a concise yet comprehensive account of the development of the mine-site concept. He emphasized that some of the figures used were simply illustrative and that even those concerned with the important parameters would change with time. Bearing in mind all the commercial factors, including the requirements of investors, the minimum period for a nodule mining operation must be at least twenty years. He pointed out that durations in excess of this period for the purposes of calculating mine-site area requirements were not justified. An annual production of three million dry metric tons a year was assumed as the basis for the calculations. Pasho also examined the 'site factors' which combined to determine the tonnage of nodules available for recovery from a mine-site and the 'system factors' which determined the proportion of the available nodules that might actually be recovered.

In his discussion of the 'site factors', Pasho emphasized that a distinction should be made between the average abundance of nodules over the whole mine-site and the average abundance over the part that would, in fact, be minable. The evolution of the ideas of the incipient industry about the characteristics and, therefore, the size of mine-sites was evidenced in his extensive references to the literature.

Assuming that a total of 4.25 million sq. km. of sea-bed contained nodules that are potentially recoverable, Pasho provided an estimate of the number of viable mine-sites associated with various methods for determining the required mine-site areas. The approach used differed significantly from earlier attempts by taking account of the likely variations in the size of mine-sites so that each could satisfy the same production factors.

2. Summary of the Discussion

The discussion on the mine-site concept and estimates of the number of mine-sites focused on the various factors that were needed to determine areal requirements. At an early stage of the discussion, it was agreed that the definitions used in Archer's paper were useful. The Group also agreed, however, that only a low level of confidence could be attached to the estimates it provided. In part of the discussion, it was made patently clear that these factors, which included water depth, topography, grade, abundance, the efficiency of the collector (dredge efficiency), and the overall maneuverability of the mining ship or ships, could not be considered in isolation but only in association. Of these complex factors it was agreed, however, that economically, the most significant variable was grade.

For a given mine-site, the opinion of the Group of Experts was that delineation of the mine-site would have taken place prior to exploitation. This would result in knowledge of such factors as the mineable portion and the abundance within this portion, perhaps with a degree of confidence of ±10%. As far as mining itself was concerned, it was pointed out that in certain cases the topography could be more important than abundance per se. The reasoning was that changes in elevation required increases or decreases in the length of pipeline, which would, among other things, present greater technical difficulties. Relatively steep slopes could increase the difficulty of guiding the sea-bed collector. The turning radius of the ships and the actual process of turning were other factors deemed important in the probable mining patterns. The total effect of all of the above made it difficult to discuss the actual areal specifications of mine-sites in the abstract.

The overall discussion suggested that no single figure could be assigned to any of the parameters. However, the Group of Experts did not express disagreement with the suggestion of the following as probable ranges for deep-sea-bed mining systems:

Factors	**Values**
Minable portion	60% to 90%
Average abundance	6500-10,000 mt/km^2 (dry)
Average grade, combined nickel and copper	2.25%
Dredge efficiency	30% to 80%
Sweep efficiency	40% to 75%

It was emphasized that these ranges could only be used subject to the explicit condition that the values were treated as being interdependent and under certain circumstances could fall outside the ranges suggested. The overall efficiency of mining systems could be regarded as likely to fall within the range 10-40% subject to similar conditions.

CHAPTER VII

RESOURCES AND POTENTIAL RESERVES OF NICKEL AND COPPER IN MANGANESE NODULES

1. Definitions

Concentrations of an element (usually combined with other elements) well above its crustal abundance can be described as *deposits* of that element or mineral. Mineral deposits that are likely to be economically workable at some time in the future are defined as *resources*: these include both those that are known and those that are not known but whose existence can be deduced. This widely accepted definition of resources immediately raises the difficult problem of deciding what will be the maximum price that will be acceptable at some time in the future, as this will determine the limit beyond which deposits will never be worked. Finally, the generally small part of resources that are economically workable in the current, locally prevailing, economic circumstances are *reserves*. It follows from this definition that resources can move into and out of the reserve category with changes in price of the commodities being produced and with changing costs of producing them. The costs may change with the development of new mining, transport and processing technology.

The distinction between these categories is of fundamental importance when the availability of any mineral raw material is being considered. Thus the total nickel and copper content of all manganese nodule deposits has no more economic significance than the corresponding figure for the total copper content of the outer part of the earth's crust.

A *mine-site* may be defined as an area in which there are sufficient nodules with a high enough grade and abundance (or density; the weight of nodules per unit area) to sustain a commerically viable mining operation. Economies of scale, particularly in the processing plant, are likely to be important and a typical operation may require about 3 million (dry)[1] tons of nodules a year for at least 20 years. Present evidence suggests that the winning of metals from manganese nodules is likely to be economically viable with the *first generation* of mining and processing equipment only if the average combined Ni and Cu content is about 2.3%, in areas where the average abundance of the nodules is about 10 (wet) kilogrammes per square metre (kg/m^2)[2]. An area from which about 60 million tons of such nodules can be recovered may be described as a *first generation mine-site*. In anticipation of the successful completion of research and development programmes, the nodules in such sites may be described as *potential reserves*, although strictly they should be described only as resources until commercial viability has been established.

It follows from the definition of reserves that the specification for a typical mine-site will change with economic circumstances. Lower unit costs, resulting from improvements in marine mining and nodule processing technology, and increases in metal prices will enable nodule deposits with lower grades and abundances to be mined in the future. Lower abundances would call for larger mine-sites to provide the same total tonnage of nodules. Larger annual capacities may be needed, but the proportion of the nodules in the mine-site that are brought to the surface will increase with improvements in mining techniques.

It is unlikely that there will be a clear distinction between first generation and later mining systems, or between other present and future factors affecting deep-sea mining, but rather a gradual evolutionary development. Nevertheless, it is useful to consider separately the potential reserves (involving prediction of the economics of first generation mining systems) and the total resources (involving very long term economic predictions).

2. The Data and the Main Assumptions

The evidence needed to attempt to estimate the resources, including potential reserves, of metals in nodules (or the corresponding numbers of mine-sites) is still sparse and imperfect. Chemical analyses from 1523 sample stations (Table 1) were accessible in the Data Bank of the Scripps Institution of Oceanography in June 1977, an average of perhaps only one station per 35,000 km of the total area in which nodules may be present.[2] Data on the world-wide variation in nodule abundance[2 3] also provide only thin coverage. Furthermore, this evidence is indirect: the published information refers to *population*, the proportion of the sea bed area occupied by visible nodules. Population can be converted to abundance only if the size of the nodules is known: in the absence of this information, the size must be assumed.[4] Unfortunately, there is also little published data relating to the extent to which acceptable grade and abundance coincide, so that this critical information must be inferred. Most of the evidence suggests that grade and abundance vary independently, although Menard and Frazer[8] have suggested an inverse relationship.[2 6 7]

In short, only a low degree of confidence can be attached to estimates based on such inadequate data. Nevertheless the data can be used, bearing in mind that there is increasing understanding of the environment in which potentially workable nodules are formed. To provide estimates at the inferred level of confidence it is assumed that:

(a) the sampling is randon;

(b) the quality of the information is equally reliable (or at least that errors are not systematic;

(c) abundance and grade vary independently.

3. Potential Reserves

For the data from all oceans in Table 1 there is a geometrical decrease in the cumulative number of samples for a unit increase in the average grade (if nodules with the lowest grades, that is, less than 0.25% combined Ni and Cu are excluded) following the relationship suggested by Lasky[10] for some deposits on land. On the basis of the assumptions above, the tonnage of nodules above any minimum grade will be proportional to the number of samples above that grade. That is, for a small increase in grade there will be a larger decrease in the tonnage of nodules available.[11]

The required average grade could, theoretically, be obtained by mining some

TABLE 1.
Summary of Data in Scripps Data Bank (June 1977): Nickel Plus Copper

Nickel plus Copper %	Number of stations (percentage of totals in brackets) Atlantic Ocean* North	Atlantic Ocean* South	Atlantic Ocean* All	Indian Ocean*	South Pacific Ocean*	North Pacific Ocean Clarion-Clipperton	North Pacific Ocean Rest	North Pacific Ocean All	All oceans All	All oceans Except Clarion-Clipperton
0 —0.25	15 (16)	29 (22)	44 (20)	22 (16)	21 (5)	1 (0)	23 (5)	24 (3)	111 (7)	110 (8)
0.25—0.50	45 (47)	50 (39)	95 (42)	52 (37)	123 (29)	1 (0)	79 (16)	80 (11)	350 (23)	349 (27)
0.50—0.75	25 (26)	23 (18)	48 (21)	31 (22)	128 (30)	2 (1)	79 (16)	81 (11)	288 (19)	286 (22)
0.75—1.00	8 (8)	15 (12)	23 (10)	13 (9)	58 (13)	11 (5)	72 (14)	83 (11)	177 (12)	166 (13)
1.00—1.25	1 (1)	6 (5)	7 (3)	14 (10)	32 (7)	9 (4)	49 (10)	58 (8)	111 (7)	102 (8)
1.25—1.50	—	4 (3)	4 (2)	2 (1)	21 (5)	21 (10)	29 (6)	50 (7)	77 (5)	56 (4)
1.50—1.75	—	—	—	4 (3)	21 (5)	15 (7)	36 (7)	51 (7)	76 (5)	61 (5)
1.75—2.00	1 (1)	1 (1)	2 (1)	1 (1)	11 (3)	16 (7)	61 (12)	77 (11)	91 (6)	75 (6)
2.00—2.25	—	1 (1)	1 (0)	1 (1)	8 (2)	30 (14)	28 (6)	58 (8)	68 (4)	38 (3)
2.25—2.50	—	—	—	—	4 (1)	32 (14)	26 (5)	58 (8)	62 (4)	30 (2)
2.50—2.75	—	—	—	1 (1)	3 (1)	37 (17)	12 (2)	49 (7)	53 (3)	16 (1)
2.75—3.00	—	—	—	—	—	28 (13)	11 (2)	39 (5)	39 (3)	11 (1)
3.00—3.25	—	—	—	—	—	14 (6)	1 (0)	15 (2)	15 (1)	1 (0)
3.25—3.50	—	—	—	—	—	3 (1)	1 (0)	4 (1)	4 (0)	1 (0)
3.50—3.75	—	—	—	—	—	1 (0)	—	1 (0)	1 (0)	—
Totals	95	129	224	141	430	221	507	728	1523	1302

Note: not all percentages add up to 100% due to rounding.

*The boundaries between Indian and Atlantic Oceans are taken at 30° E, between Indian and South Pacific at 150° E and between South Pacific and Atlantic at 70° W.

nodules with a grade well above the average and something containing very little nickel and copper. In practice, as with ore bodies on land, a minimum or cutoff grade will be calculated: a combined Ni, Cu and Co content of 2% has been suggested (Flipse et al.). If the cobalt is ignored the cutoff can be expressed as 1.76% combined Ni and Cu.[12] This grade is exceeded in only 15% of the samples according to Healing and Archer, and, therefore, in only about 15% of all nodules.

Nodules are recorded at only about 15% of the camera stations which are the basis of the world-wide nodule population data (Table 2). This suggests that nodules are present in a total area of about 54 million km^2 with, coincidentally, an average abundance of about 10 kg/m^2 (assuming that the average diameter is 3.68cm), the requirement for first generation mine-sites. However, although small areas within mine-sites with a very low abundance (or no nodules) may be mined, larger areas with less than a minimum, or cutoff, abundance normally would be avoided. If a cutoff of 5 kg/m^2 is assumed,[13] the total area would be reduced by about 40% to about 32.5 million km^2.

Thus, considering the minimum grade and abundance together, this method suggests that 4.9 million km^2 meet the specification for first generation mine-sites (or 4.4 million km^2 if the average diameter is 2.54 cm).

TABLE 2.
Population Distribution[2]

	Number of stations (percent of total in brackets)				
	Atlantic Ocean	Indian Ocean	South Pacific	North Pacific	Total
No. of stations	1104	271	378	557	2310
No nodules	1009 (91)	218 (30)	295 (78)	433 (78)	1955 (85)
With nodules	95 (9)	53 (20)	83 (22)	124 (22)	355 (15)
Less than 25%	49 (51)	27 (51)	35 (42)	57 (46)	168 (47)
25 to 50%	16 (17)	13 (25)	21 (25)	24 (19)	74 (21)
50 to 75%	13 (14)	7 (13)	16 (19)	21 (17)	57 (16)
More than 75%	17 (18)	6 (11)	11 (13)	22 (18)	56 (16)

The particular criticism of this method is that, because it deals with world-wide average grade and abundance, the total will include small areas which, while meeting the specification, are so widely dispersed that they could not be mined. Furthermore, part of the total will consist of areas, some probably large, in which the grade or the abundance (or both) are above the cutoff but below the specified average; some, at least, of the areas in which the grade or abundance is correspondingly above average will be too far away (even in a different ocean) to be blended to produce the required averages. The estimated total will therefore be too high and serves only as an extreme upper limit.

An alternative method is to attempt to estimate the extent of *"prime areas"*, which may be defined as areas in at least part of which there are deposits of relatively abundant nodules with significantly higher grades than elsewhere. They are the areas in which mine-sites are likely to be found. Comparison of maps showing the distribution of grade[14] [15] and population (Skornyakova, Andrushchenko, Ewing et al. previously cited) indicates that only a small proportion of the sea-bed is likely to satisfy this definition. This evidence points clearly to the largest prime area being in the North Pacific between the Clarion and Clipperton fracture zones, about 3.4 million km^2, from about 7° to 15° N and 120° W to 155° W. Another area (about 0.8 million km^2) in the North Pacific, to the southwest of Hawaii, also qualifies. Although a mirror image of the Clarion-Clipperton prime area might be expected to be present in the South Pacific, the area in a corresponding position relative to the Equatorial zone of high productivity is restricted by the presence of part of the East Pacific Rise and the Marquesas and Tuamotu archipelagoes. There may be about 1 million km^2 of prime area in the South Pacific. Some high grade nodules have been recovered from the floor of the Indian Ocean and if an additional, say, 15% is allowed for the Atlantic Ocean and for areas that have not yet been sampled, the total prime area might be very approximately 6.5 million km^2 (Table 3). This is less than 2% of the ocean floor.

TABLE 3.
Estimated Prime Areas

	million km^2
North Pacific Ocean	4.2
South Pacific Ocean	1.0
Indian Ocean	0.5
Subtotal	5.7
Atlantic Ocean and unexplored areas	0.85
Total (approximately)	6.5

Only part of the total prime area will meet the specification for first generation mine sites. The same principles can be applied to the Clarion-Clipperton prime area as were used for the whole of the ocean floor in the first method described above, although with slightly more confidence as the average sampling interval is closer (about 1 station to 15,000 km^2). Of the 221 samples from this area (Table 1) 161, that is 73%, are above a cutoff of 1.76% combined Ni and Cu. Abundance data for the western half of the Clarion-Clipperton prime area (obtained from stations spaced at an average of about 1 to 11,800 km^2) suggest, assuming that the nodules have an average diameter of 3.68 cm, that about 42% has more than a minimum abundance of 5 kg/m^2. Combining grade and abundance the proportion is about 30%. However, Schultze-Westrum's data, while supporting the assumption that grade and abundance are independent variables, also suggest that the specified grade and abundance coincide in about 38% of the same area, that is 11 out of 29 stations. As a rough approximation it might be assumed that one-third of the Clarion-Clipperton prime area meets the requirement for first generation mine sites.

If it is assumed that the proportion of all prime areas with acceptable grade and abundance is the same as in the Clarion-Clipperton area, then the total first generation mine-site area would be very approximately 2.2 million km^2.

However, this method is also open to the objection that not all of this total area would sustain first generation mining operations, as it would include areas with the necessary characteristics but too small to be mined and larger areas below the required average grade or abundance that are too widely dispersed to be blended. But as only prime areas are considered, rather than the whole ocean floor, such deposits will account for a smaller proportion and the estimate should be nearer the practical figure than that provided by the first method, if the estimate of the total prime area is reasonable.

Healing and Archer found that above a cutoff grade of 1.76% the average combined nickel and copper content would be about 2.29%. Above a cutoff abundance of 5 kg/m^2 the average may be about 15 kg/m^2 for nodules with an average diameter of 3.68cm (and about 12 kg/m^2 if it is 2.54cm). The effect of the choice of some variables on the possible maximum number of first generation mining operations that might be sustained by these estimated total areas is illustrated in Table 4. This involves a further assumption, the level of the overall mining recovery efficiency. The overall mining recovery efficiency takes account of the proportion of the mine-site that cannot be mined because of physical obstructions and grade or abundance below the required level, the proportion that will be swept and the proportion of the nodules in the swept area that will be delivered to the ship. For first generation mining equipment this is likely to be between 10% and 50%, but seems more likely to be between 20% and 25%.

The inferred potential reserves of metals in the total first generation mine-site areas estimated by the two methods are shown in Table 5. These refer to quantities of contained metal, *in situ*. (First generation mining recovery from the sea bed is likely to be substantially lower than recovery on land; processing losses are likely to be similar.)

TABLE 4
Estimates of Maximum Number of First Generation Mining Operations*, Showing the Effect of Different Assumptions

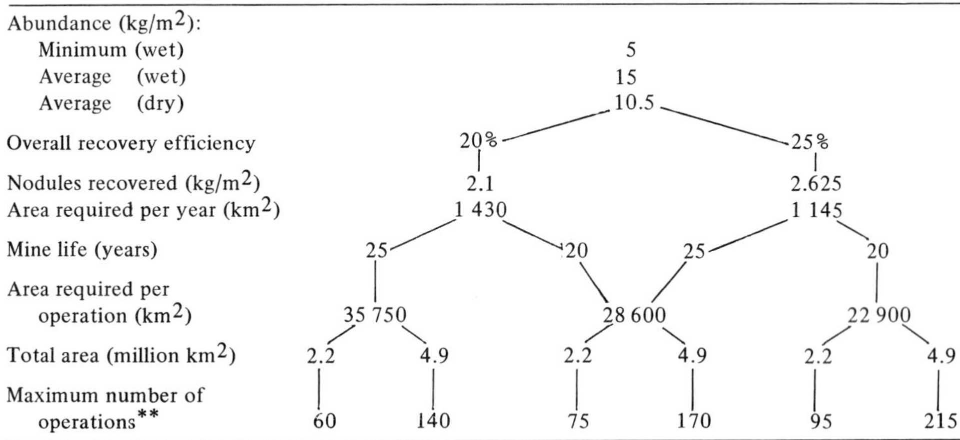

Abundance (kg/m^2):						
Minimum (wet)	5					
Average (wet)	15					
Average (dry)	10.5					
Overall recovery efficiency	20%				25%	
Nodules recovered (kg/m^2)	2.1				2.625	
Area required per year (km^2)	1 430				1 145	
Mine life (years)	25		20 / 25		20	
Area required per operation (km^2)	35 750		28 600		22 900	
Total area (million km^2)	2.2	4.9	2.2	4.9	2.2	4.9
Maximum number of operations**	60	140	75	170	95	215

*Assuming each has a throughput of 3 million (dry) tonnes a year and that the average diameter of the nodules is 3.68 cm (1.45 in).

**The total area available seems likely to be closer to 2.2 than 4.9 million km^2 (see text): the number of operations is therefore likely to be closer to the lower estimate in each of the three pairs. Figures rounded to nearest five.

TABLE 5
Potential Reserves and Resources in Nodules

	Potential Reserves	Resources
Assumed minimum grade (%Ni + Cu)	1.76%	0.88%
Average grade (%Ni + Cu)	2.29%	1.57%
Assumed minimum abundance (wet kg/m²)	5	2.5
Assumed average nodule diameter[2] (cm)	3.68	3.68
Average abundance (wet kg/m²)	15	13.5
Estimated area with this grade and abundance (wet kg/m²)	2.2[1]	18.5
Total nodules (wet million tons)[4]	33,000	250,000
Total nodules (dry million tons)[4]	23,000	175,000
Assumed average Ni grade	1.26%	0.86%
Ni content (million tons)[3]	290	1,500
Assumed average Cu grade	1.03%	0.71%
Cu content (million tons)[3]	240	1,240
Assumed average Co grade	0.25%	?
Co content (million tons)[3]	60	?
Assumed average Mn grade	27.5	?
Mn content (million tons)[3]	6,000	?

[1]See footnote 2 to Table 4.
[2]All tonnages will be less if the average diameter of the nodules is less than 3.68 cm (1.45 in): for example, they would be reduced by about 30% if the averge diameter is 2.54 cm (1 in).
[3]Maximum quantities *in situ* rounded to nearest 10 million tons (manganese rounded to nearest billion (10^9): recoverable metal would be substantially less (see paras. 12, 16 and 24: there would also be a processing loss).
[4]Rounded to nearest billion (10^9).

Very considerable uncertainty is attached to the estimates suggested in Tables 4 and 5, because of the inadequacy of the data used and doubts about the assumptions that must be made. For example, the specification assumed for a commercially viable mine-site is based on estimates of costs, future prices and recoveries, all of which may be proved wrong. Nevertheless, it is encouraging that the estimates provided by the two methods are not greatly dissimilar. They are consistent with those obtained by the same two methods are not greatly dissimilar. They are consistent with those obtained by the same two methods but with different choices of variables.[17]

4. Resources

The difficulties attending the estimation of resources are even greater than those associated with reserves, for an attempt must be made to predict the grade and abundance which will be workable in the distant future. The following section is, therefore, highly speculative.

For the purpose of discussion, the eventual lower limits of workable grade and abundance are assumed to be 0.88% combined Ni and Cu and 2.5 kg/m² (about ½ lb/ft²), that is half the cutoffs assumed for potential reserves.

The relationship between grade and tonnage, inferred from the relationship between grade and number of samples by Healing, Archer and Frazer, suggests that there may be about three times (45%) as many nodules, of those with at least 0.25%

combined Ni and Cu, with more than 0.88% combined Ni and Cu as there are with more than 1.76% (15%: see Figure 1). At the same time the tonnage of combined nickel and copper that they contain may only double, from 33% of the total to 68% (Figure 1). That is, if abundance is ignored the potential reserves represent about half of the total resources above the arbitrarily selected cutoff grade for resources.[19] The average grade may be about 1.57% combined nickel and copper.

FIGURE 1. All Oceans: Cumulative Percentages of Nodules and Metals Against Grade

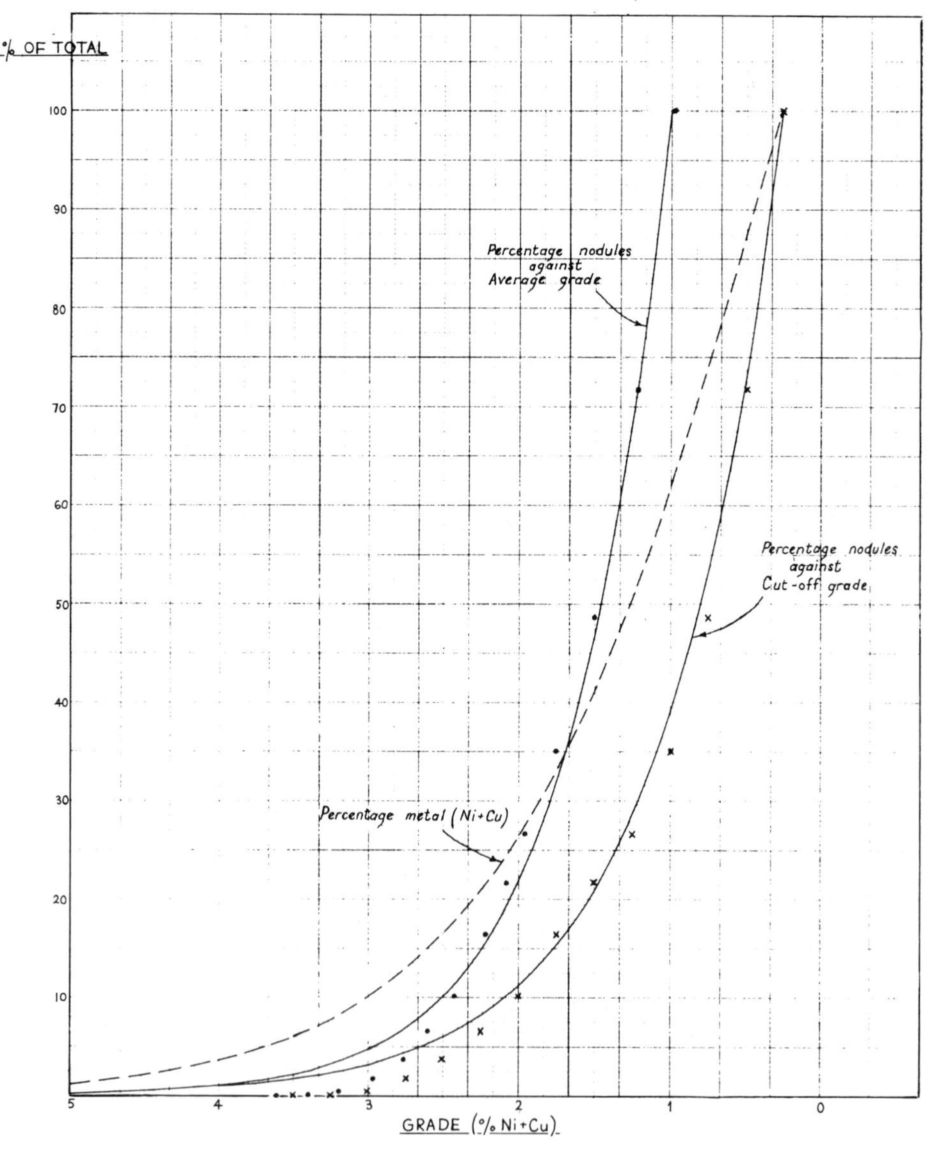

The data on nodule population distribution indicates that about 41 million km^2 of the ocean floor may be occupied by nodules with an abundance of at least 2.5 kg/m^2. As about 45% of this area may have nodules above the cutoff grade, the area with nodules above both the required minimum grade and abundance may therefore be about 18.5 million km^2. This area would contain about 175,000 million tons of nodules.[21]

However, a considerable proportion of these nodules is likely to be unworkable, even eventually, being too widely dispersed in areas that are either too small or that fail to meet the combined abundance and grade criteria since, as with one estimate of potential reserves, the figure of 175,000 million tons is based on ocean-wide data. Therefore, this estimate should be regarded only as a theoretical (and very approximate) maximum.

Increases in the proportion of the nodules that can be recovered from the ocean floor are likely to be a major factor encouraging the exploitation of deposits of lower grade and abundance. It may be reasonable to speculate that overall mining recovery efficiency will not improve beyond about three times that assumed for first generation mining equipment, that is to about two-thirds (67%). However, this is likely to be achieved only in the distant future, so that the average recovery achieved by all mining systems may be only, say 45%. The total quantity of nodules that could be economically recovered, eventually, would then be considerably less than a theoretical maximum of about 78,750 million tons.

Furthermore, annual throughputs of more than 3 million tonnes may well be essential at lower grades and abundances. It is thus impossible to estimate, even roughly, the ultimate number of mining operations that will prove possible. Clearly, there will be considerably more than the number of first generation mine-sites and very many less than the theoretical maximum of about 1300 operations, even if allowance is made for the possibility that it will prove economical for first generation sites to be reworked by advanced mining systems with a high overall recovery efficiency.

5. Conclusions and Their Inaccuracies

It cannot be emphasized too strongly that all these estimates remain very uncertain. They are based on the inadequate data that are available at present and involve a number of sweeping assumptions that remain to be tested. For example, revision, from 13% to 15%, of the proportion of nodules with the grade assumed to be required for first generation mines (on the basis of data from substantially more stations) and revision of the assumed minimum abundance requirement, from 10 kg/m^2 to 5 kg/m^2, have led to amendment of the estimates of the number of first generation mine-sites (Table 4) of as much as 38% above those given in previous essays.[18,22]

In particular, by both methods the proportions of nodule deposits in very large areas of the sea bed that meet the assumed minimum grade and abundance criteria are first considered independently. These are then combined arithmetically on the assumption that they are independent variables, which introduces the 'dispersal' effect (paras. 12, 16, and 24). This inevitably leads to the inclusion of deposits that will not be mined (on the basis of the same criteria). The smaller the area considered, the more this effect will be reduced, so that the prime areas method provides a more realistic estimate provided that their extent has been estimated with sufficient accuracy. This effect is minimized by determining both grade and abundance at closely spaced stations, the procedure used in detailed evaluation of an area. The dispersal effect is reduced in the estimates provided by Frazer[23] and Bastien-Thiery and others[24], as they consider smaller areas.

Nevertheless, the estimates suggested in Tables 4 and 5 may at least indicate the orders of magnitude. The estimates are not inconsistent with those offered by Pasho and McIntosh[25] using a different method and if so, it seems that the quantities of nickel, copper and manganese that would become available from manganese nodules are likely to be neither enormously greater nor enormously less than remain to be mined on land. Cobalt may be an exception.

Notes and References

[1]The custom is followed that the wet weight (i.e. saturated with water) is used for nodules *in situ*, all other figures referring to the dry weight, which is about 30% less.

[2]Skornyakova, N.S. and Andrushchenko, P.F. Iron-manganese nodules in the Pacific Ocean. *Litol. i polez. Iskop.*, 1964, No. 5, pp. 21-36. [In Russian.]

[3]Ewing, M., Horn, D., Sullivan, L., Aitken, T. and Thorndike, E. Photographing manganese nodules on the ocean floor. *Oceanology Int.*, 1971, Vol. 6, No. 12, pp. 26-27, 30-32.

[4]An abundance of 10 kg/m² is represented by a population of 35%, if the average diameter is 3.68 cm (145 in) [Ref 3], but the same population corresponds to an abundance of 7 kg/m² if the average diameter is 2.54 cm (1 in).

[5]Flipse, J.E., Dubs, M.A. and Greenwald, R.J. Pre-production manganese nodule mining activities and requirements. Pp. 602-800 *in* Mineral Resources of the Deep Seabed. Hearings before the Subcommittee on Minerals, Materials, and Fuels of the Committee on Interior and Insular Affairs, United States Senate, 93rd Congress, 1st Session on S. 1134. (Washington, D.C.: U.S. Government Printing Office, 1973).

[6]Schultz-Westrum, H.H. The station and cruise pattern of the R/V Valdivia in relation to the variability of manganese nodule occurrences. Pp. 145-149 *in Papers on The Origin and Distribution of Manganese Nodules in the Pacific and Prospects for Exploration,* an International Symposium organized by the Valdivia Manganese Exploration Group and the Hawaii Institute of Geophysics, Honolulu, July 1973. Morgenstein, M. (Editor).

[7]Schatz, C.E. Observations of sampling and occurrence of manganese nodules. Paper OTC 1364, 1971. Offshore Technology Conference, Houston, Tex., Preprints, Vol. 1, pp. 389-396.

[8]Menard, H.W. and Frazer, J.Z. Inverse correlation between grade and abundance of manganese nodules. *Mar. Min. In press.*

[9]Healing, R.A., Frazer, J.Z. and Archer, A.A. The frequency distribution of nickel, copper and cobalt grades in manganese nodule deposits 1977. 12 pp. [Unpublished manuscript.]

[10]Lasky, S.G. How tonnage and grade relations help predict ore reserves. *Engng Min. Jnl.*, 1950, Vol. 151, No. 4, pp. 81-85.

[11]For example, if the required average grade is increased by 0.1 percentage point (from, say, 2.3 to 2.4% combined Cu and Ni) the tonnage of nodules available would probably decrease by about 14%, or the tonnage of nodules would increase by about 360% if the average grade required is decreased by 1.0 percentage point (from, say, 2.3 to 1.3%). It should be noted that the quantity of metal available will not vary by the same proportions, as the average grade has been changed.

[12]Perhaps with an average of about 0.97% Ni and 0.79% Cu. This cut-off grade can be represented as a nickel equivalent of 1.23%, (Ni + Cu), if the value of the nickel is assumed to be three times that of copper. On the same basis, an average grade of 2.29% combined Ni and Cu corresponds to a nickel equivalent 1.6%.

[13]The adoption of a minimum abundance higher than zero automatically increases the average abundance, in this case to about 15 kg/m^2.

[14]Frazer, J.Z. and Arrhenius, G. World-wide Distribution of Ferromanganese Nodules and Element Concentrations in Selected Pacific Ocean Nodules. 55 pp. (Washington, D.C.: Office for the International Decade of Ocean Exploration, National Science Foundation, 1972). Technical Report No. 2. [Unpublished manuscript.]

[15]Horn, D.R., Horn, B.M., and Delach, M.N. Metal Content of Ferromanganese Deposits of the Oceans. 51 pp. (Washington, D.C.: Office for the International Decade of Ocean Exploration, National Science Foundation, 1973.) Technical Report No. 3. [Unpublished manuscript.]

[16]Arrhenius, G. *in* Holser, A.F. [Ref. 12].

[17]Holser, A.F. Manganese nodule resources and mine-site availability. Professional Staff Study, Ocean Mining Administration. 12 pp. (Washington, D.C.: U.S. Department of the Interior, 1976.)

[18]Archer, A.A. Prospects for the exploitation of manganese nodules: the main technical, economic and legal problems. Paper presented at I.D.O.E. Workshop, Suva, Fiji, Sept., 1975. Pp. 21-38 in *Tech. Bull. No. 2, CCOP/SOPAC,* 1976. Glasby, G.P. and Katz, H.R. (Editors).

[19]There is no general rule that the quantity of contained metal is doubled if the minimum grade is halved.

[20]By interpolation, the corresponding population (9%, assuming an average diameter of 3.68 cm) occurs over 75% of the area in which nodules may be present (15% of the whole ocean floor), so that the area is 362 million km^2 × 15% × 75%.

[21]The interpolated average abundance is 13.5 kg/m^2, so that there would be about 250,000 million wet tonnes.

[22]Archer, A.A. Economic aspects: the definition of nodule resources and their extent. Paper presented at EEC Seminar for ACP experts to UNCLOS on "The Exploitation of the Deep-Seabed", Brussels. 1977. 14 pp.

[23]Frazer, J.Z. Manganese nodule reserves: an updated estimate. *Mar. Min. In press.*

[24]Bastien-Thiry, H., Lenoble, J.P. and Rogel, P. French exploration seeks to define mineable nodule tonnages on Pacific floor. *Engng Min. Jnl.,* 1977, Vol. 178, No. 7, pp. 86-87, 171.

[25]Pasho, D.W. and McIntosh, J.A. Recoverable nickel and copper from manganese nodules in the northeast equatorial Pacific — preliminary results. *Can. Min. Metall. Bull.,* 1976, Vol. 69, No. 773, pp. 15-16.

CHAPTER VIII

DETERMINING DEEP SEA-BED MINE-SITE AREA REQUIREMENTS — A DISCUSSION

I. Introduction

1.1 Background and Purpose

"In the jargon of deep seabed mining, the term "mine-site" is widely accepted and generally used...to describe the area that will be needed to sustain a manganese nodule operation."[1] There seems, however, to be little agreement as to what would be the actual area requirements for such a mine-site. The variety of proposed mine-site sizes, coupled with occasional changes in those estimates has led some to suggest that the situation is confused to such an extent that it is virtually impossible to sort out, analyse and select from the alternatives. It is, therefore, the purpose of this paper to review existing estimates of area requirements and discuss optional approaches for establishing such requirements.

1.2 Review of Factors Influencing Area Requirements

For purposes of discussion, the factors that influence the area requirements for a deep sea-bed mine-site are grouped into (1) *production factors,* which determine the cumulative recovery required from a mine-site, (2) *site factors,* which combine to indicate the tonnage of nodules available for recovery from a mine-site and (3) *system factors,* which will determine the portion of available nodules that might actually be recovered.

The following review has benefited considerably from the works of Flipse *et al.,*[2] Kaufman[3] and Siapno[4] and it differs only slightly from their presentations.

1.2.1 Production Factors

The cumulative recovery (r_c) of nodules from a mine-site can be determined from the following:
Annual recovery rate (r_a) is simply the annual tonnage of nodules to be recovered from a mine-site. In cases where more than one mining system is to operate on a single site, it is the total annual recovery rate of all such units.

Duration (d) is the number of years that a mining unit (or group of units) is envisioned to be in production at a given recovery rate.

1.2.2 Site Factors

The characteristics of an individual mine-site and thus the quantity of nodules theoretically available for recovery will vary considerably from site to site, depending on:

Mineable proportion (m) of a site is defined as the ratio of an area in which mining will actually be attempted to the total mine-site area. Excluded from the mineable portion are those areas of a mine-site which contain topographic features, obstacles and other bottom conditions such that it would be dangerous and/or impractical to attempt recovery in their vicinity. Also excluded are zones where nodules are below cut-off grade and/or abundance.

Nodule abundance (a_m) is used herein as the average weight of manganese nodules per unit area in the mineable portion of a mine-site. Abundance is always expressed as dry metric tons of nodules (dmt) per square kilometre.

1.2.3 System factors

The characteristics of an individual mining system that will determine its effectiveness in recovering nodules from a mineable area include:

Dredge efficiency (e_d) is the ratio of the quantity of nodules recovered by a collecting device to the quantity of nodules available in the dredge path.

Sweep efficiency (e_s) is the ratio of the area actually swept by the collector vehicle to the mineable area.

1.3 Calculating Formulae

For a specific sea-bed mining operation, mine-site size can be approximated by

$$A_s = \frac{r_a d}{m \; a_m \; e_d \; e_s}$$

where:

A_s= total mine-site area (km^2)

m = proportion of a minesite that is mineable; ratio of the area in which mining would be attempted to the total mine-site area

r_a = annual nodule recovery rate (dmt/year)

d = duration of the operation (years)

a_m= average nodule abundance in the mineable portions of a minesite (dmt/km^2)

e_d = dredge efficiency

e_s = sweep efficiency

Because the individual parameters for equation (1) are commonly unspecified, two additional factors are defined in order to facilitate the evaluation and comparison of mine-site size estimates.

Net nodule recovery ratio (R) denotes the average quantity of nodules recovered per unit of mine-site area and is given by $R = \frac{r_a d}{A_s}$

or, substituting for A_s from equation (1). $R = m \; a_m \; e_d \; e_s$

Note that net nodule recovery as expressed in equation (3) is independent of the recovery rate, duration and area factors.

Net mining efficiency (E), is the percentage of nodules in the total mine-site area, A_s, that is actually recovered and is given by: $E = \frac{r_a d}{A_s \; a_s}$

or, since $r_a \cdot d = A_s \cdot m \; a_m \; e_s \; e_d$ (from equation 1),

$$E = \frac{R}{a_s}$$

where:

a_s = average nodule abundance in the total mine-site area, A_s.

As a_s and a_m are rarely distinguished, computation of E assuming $a_s = a_m$, yields

$E = m \;\; e_d \;\; e_s$ (6)

E has also been termed "overall mining recovery efficiency" and "mining efficiency" in the literature. The approximation in equation (6) appears widely in the literature. The possible errors resulting from the approximation are discussed in a later section of this paper.

Thus, for example, a mining operation that is to recover three million dmt/year for 20 years from a mine-site area that is 75% mineable (m = .75), has an average abundance in mineable portions of 6,5000 dmt/km^2, using a system that has a dredge efficiency of .50 and a sweep efficiency of .65 would require a total mine-site of approximately 38,000 km^2. Net nodule recovery would be roughly 1,600 dmt/km^2 and net mining efficiency is approximately .25.

2. Historical Review

2.1 Early Suggestions

Even the earliest discussions related to mine-site area requirements recognize the unique and variable nature of nodule deposits and the difficulty of assigning a universally applicable mine-site size. Brooks[5], for instance, suggested that the size not be fixed, but rather be variable and based upon the character of the resource in the area and the technology available for its recovery. Christy,[6] while indicating an adequate site size might be at least 2,600 km^2,[7] noted that fixing the size of all sites would result in considerable variation in their value because of differences in, for example, abundance, grade and water depth.

One of the earliest proposals that can be analysed is that of Mero.[8] He suggests a "claim" size of 13,000 km^2, which would contain sufficient material to support a 4.5 million dmt/year operation for 20 years (cumulative recovery approximately 91 million dmt, yielding a net nodule recovery of 7,000 dmt/km^2). Use of 4.5 million dmt/year is approximately 4 times the annual recovery rate that he employs in his "standard" sea-bed mining operation, and it would seem that he envisions several mining systems operating on the same "claim". A figure of 7,000 dmt/km^2 for net nodule recovery is consistent with his other works that assume an abundance of 9,800 dmt/km^2 and a dredge efficiency of about .7 (Mero, 1965). As suggested by his calculations and discussions, it could be concluded that the implications of topography, the characteristics of the joint distribution of grade and abundance, and the sweep efficiency were either not recognized or not appreciated. There is also some ambiguity about the units (wet or dry weight) used for his abundance figures; since his analyses are given on a dry weight basis, it would seem as if all units were intended to be dry weights.

It is interesting to note that information stemming from academic research provided preliminary indications that problems might arise due to variations in abundance.[9]

Industry probably could not have provided much more information on site factors at that time. Newport News Shipbuilding and Drydock Company (predecessor of Deepsea Ventures Inc.) had concentrated its earlier exploration work on the shallow water nodules of the Blake Plateau and Pacific seamounts, while Kennecott Copper Corp. (Bear Creek Mining Co.), having conducted a single cruise some 5 years earlier, indicated that insufficient deep ocean data had been gathered to even establish that the nodule deposits constituted a resource.[10]

2.2 Deepsea Ventures Inc.[11]

Deepsea Ventures' rather unusual policy of periodic disclosure of information regarding their ocean mining project has generated sufficient historic information to allow analysis of the evolution of the company's opinion on the subject of mine-site size (Table 1).

TABLE 1.
Comparison of Minesite Area Requirements — Deepsea Ventures Inc.*

	1969-70	1971		1974: "Deposit"		
Minesite Area Required (km²)	2,600	10,000	30,000	30,000	30,000	30,000
Production Factors						
Recovery (x 10⁶ dmt/year)	.9	.9	.9	.9 for 35 years,	.9 for 20 years,	2.7
Duration (x 10⁶ dmt)	20	20	40	2.7 for five years	2.7 for 20 years	40
Cumulative Recover (x 10⁶ dmt)	18.1	18.1	36.3	45.0	71.9	107.5
Site Factors						
Mineable				.75**		
Abundance (dmt/km²)	9,750	6,500	6,500	6,500	6,500	6,500
System Factors						
Dredge Efficiency				.50		
Sweep Efficiency				.65		
Net Nodule Recovery (dmt/km²)	7,000	1,800	1,200	1,500	2,400	3,600
Net Mining Efficiency	.70	.25	.20	.25	.35	.55

*all factors may not arithmetically agree due to rounding.
**takes into account areas below cut-off grade and abundance.

In 1969 and 1970, various statements indicated that a lease area of 2,600 km² would be needed in order to produce approximately .91 million dmt/year for 20 years.[12] Their anticipated net nodule recovery was in the order of 7,000 dmt/km², and, using a required average abundance of 8,000 to 9,800 dmt/km², net mining efficiencies of .72-.86 were apparently envisioned.

Such high net mining efficiencies imply a mine-site that is virtually entirely mineable and/or a mining system with extremely high dredge and sweep efficiencies (see equation 6). Recognizing that topography was of "vital concern," they apparently believed that they could locate "topographically favourable" areas of sufficient size. High dredge efficiencies may reflect an optimistic interpretation of the results of dredge head tests conducted in test facilities by Deepsea's predecessor (Newport News Shipbuilding and Drydock Co.) in the early 1960's. High sweep efficiencies may have been assumed for a mining system characterized by accurate dredge head control. It is interesting to note that, although Deepsea's area requirements were roughly 20 percent of Mero's, their net nodule recovery and net mining efficiency are virtually identical, the variation resulting only from the use of different recovery rates. The similarity may well reflect a common appreciation gained during Newport News' earlier nodule project with which both Deepsea personnel and Mero were associated.

By mid-1971, while still envisioning a .91 million dmt/year operation for 20 years, Deepsea Ventures increased their estimated area requirements from 2,600 km^2 to slightly over 10,000 km^2 [13] thus decreasing net nodule recovery by roughly 75% from 7,000 dmt/km^2 to 1,800 dmt/km^2. Commencing their first deep sea-bed exploration in the North Pacific during mid-1969, they shortly thereafter located the area that was to constitute their "Deposit"[14] and subsequently announced in mid-1971, that they were "...ready to file a claim on a specific orebody now."[15] Because most of Deepsea Ventures' survey work from 1971 through 1973 consisted of joint or charter cruises conducted throughout the northeast equatorial Pacific, the twelve to sixteen months of survey time spent defining the extent of their "Deposit" prior to 1974 was probably done between late 1969 and mid-1971. If so, the reassessment of site factors in 1971 largely reflects the characteristics of their specific mine-site. In addition to the new site information, prototype mining tests conducted in approximately 800 metres water depth off the east coast of the United States during 1970 provided better information from which to estimate system factors.

Thus the increase in mine-site area requirements indicated by Deepsea Ventures in 1971 likely reflects a reassessment of site factors specific to an area they were prepared to claim, and system factors re-evaluated in light of prototype mining tests.

The only subsequent change in mine-site size requirements was announced in association with Deepsea Ventures' 1974 "Notice of Discovery and Claim of Exclusive Mining Rights...", in which they said that they plan to retain a 30,000 km^2 mine-site from which they would recover between .91 million and 2.7 million dmt/year for as long as 40 years. Information released by Deepsea Ventures that might provide a basis for analysing this increase in mine-site area requirements is sufficiently confusing so as to allow only a rather subjective discussion of possible alternatives. Certainly, one of the major reasons for the increase in area requirements is the provision allowing for 40 years of production as opposed to the previously envisioned 20. One of the first indications that Deepsea Ventures intended to increase the duration of its operation came in 1972, when it was stated that "Candidacy (mine-site) is judged by the requirements of a 40-year operation at (approximately .91 million metric tons) of dry nodules recovered per year.[16] Subsequently, in 1973, Deepsea Ventures indicated that it would be difficult to meet their cumulative recovery requirements (.91 million dmt/year for 40 years) from 10,000 km^2 [17] and eventually noted in their "claim" of 1974 that provision had indeed been made for a duration of up to 40 years.

A portion of the increase may be attributable to the allowance that Deepsea Ventures makes for increasing production from their historic standard of .91 million dmt/year to as much as 2.7 dmt/year. The real effect of providing for flexibility in production, however, is very much a function of net nodule recovery. A review of the literature relating to Deepsea Ventures' evaluation of net nodule recovery between 1971 and 1974 suggests that values in the range of 1,500 to 1,800 dmt/km^2 are not unrealistic. Net nodule recoveries approaching 1,800 dmt/km^2 are implied by their 1971 mine-site requirements of approximately 10,000 km^2. A somewhat lower value of 1,500 dmt/km^2 was utilized by the American Mining Congress in 1973 to support Deepsea's claim that it would be difficult to produce .91 million dmt/year for 40 years from 10,000 km^2.[18] In all cases abundance (a_m) is 6,500 dmt/km^2, which is the average abundance given by Deepsea Ventures for its "Deposit", and this implies net mining efficiencies in the order of .25. The apparent consistency during this period of time is not expected, as Deepsea Ventures had completed the major part of the mine-site delineation and prototype mining evaluation prior to mid-1971.

Evaluating Deepsea Ventures' "claim" of 30,000 km^2 at product rates of .91 million and 2.7 million dmt/year and at a net nodule recovery of approximately 1,500 dmt/km^2 indicates a production schedule of approximately 35 years producing at a rate of .91 million dmt/year and 5 years at 2.7 million dmt/year. Taken literally, this is not a realistic production schedule. A better interpretation would be to assume that sufficient leeway had been incorporated into the area requirements in order to allow flexibility in the event that either higher than expected cumulative production (as much as a 25% increase in production) were achievable from a system originally envisioned to recover .91 million dmt/year or a lower than expected net nodule recovery (as much as 20% lower than expected) were experienced.

In order to produce for 20 years at .91 million dmt/year and 20 years at 2.7 million dmt/year, a net nodule recovery of 2,400 dmt/year would have to be achieved (E = .37) and to produce for the full 40 years at 2.7 million dmt/year, a net nodule recovery approaching 3,600 dmt/year would be necessary (E = .55). As several sources suggest that net mining efficiencies near .50 may be achieved over the long term (see subsequent discussion of mining efficiency estimates), the mine-site area requirement would allow the company to take full advantage of such developments, and thus again incorporates considerable flexibility.

In summary, it would appear that the increase fom 10,000 km^2 to 30,000 km^2 results principally from the doubling of duration, and secondarily from allowing for sufficient flexibility to accommodate efficiencies lower than anticipated as well as providing for the possibility of significant production expansion in the event higher mining efficiencies were achieved.

2.3 American Mining Congress

While Deepsea Ventures Inc. has both individually, and through the American Mining Congress (AMC), advanced its opinions on mine-site size, others such as Kennecott Corp. have almost exclusively utilized the AMC to put forward its view on this matter. The United States Deep Seabed Hard Minerals Resource Act (S.2801), first considered in late 1971 and drafted in part by the AMC, contained a provision for a retained mine-site area of 10,000 km^2. Originally, the AMC supported the provision for a 10,000 km^2 mine-site size. In 1973, however, AMC submitted information to the Chairman of the United States Senate Subcommittee on Minerals, Materials and Fuels indicating that they had "...concluded that the 10,000 square-kilometre block size is in fact barely adequate and that multiple blocks will be required for some operators."[19]

The AMC's background material was based upon information presented in a paper prepared by Deepsea Ventures Inc. and Kennecott Copper Corp. This paper, reviewed factors that influence mine-site size and provided estimated ranges for these factors. In support of their contention that 10,000 km² is insufficient for many operations, they cited examples using 40 years duration and recoveries of .91 million and 2.7 million dmt/year which indicate area requirements of approximately 24,500 km² and 73,000 km². Both examples yield a net nodule recovery of roughly 1,500 dmt/km² and net mining efficiencies of slightly less than .25 (Table 2).

TABLE 2.
Comparison of Minesite Area Requirements — American Mining Congress*

	1973		1976			
			"with anticipated efficiency"		"best conditions of efficiency"	
Minesite Area Required (km²)	25,000	73,000	22,000	66,000	16,000	46,000
Production Factors						
Recovery (x 10^6 dmt/year)	.9	2.7	.9	2.7	.9	2.7
Duration (x 10^6 dmt)	40	40	40	40	40	40
Cumulative Recovery (x 10^6 dmt)	36.3	107.5	36.3	107.5	36.3	107.5
Site Factors						
Mineable	.7**	.7**				
Abundance (dmt/km²)	6,500	6,500	6,500	6,500	6,500	6,500
System Factors						
Dredge Efficiency	.50	.50				
Sweep Efficiency	.65	.65				
Net Nodule Recovery (dmt/kn²)	1,500	1,500	1,600	1,600	2,300	2,300
Net Mining Efficiency	.25	.25	.25	.25	.35	.35

*factors may not arithmetically agree due to rounding.
**takes into account areas below cutoff grade and abundance.

In 1976, the AMC suggested that area provisions in U.S. Deep Seabed Mineral legislation be amended to provide for a mine-site size of 30,000 km². Using the same calculating formula as before, they provide examples of area requirements associated with mining operations of 40 years duration and annual recoveries of .91 million and 2.7 million dmt/year for "anticipated efficiency" and "the best conditions of efficiency."[20] The results, summarized in Table 2, when analysed assuming an average abundance (a_m) of 6,500 dmt/km², indicate "anticipated efficiency" and "best conditions of efficiency" would correspond to net nodule recoveries of approximately 1,600 dmt/km² and 2,300 dmt/km² with net mining efficiencies of roughly .25 and .35 respectively.

It was indicated by the American Mining Congress that the increase from 10,000 km² to 30,000 km² was due to the fact "...that since 1970, ..., it has become increaseingly clear, especially as technological data have been accumulated, that the residual (e.g. after relinquishment) 10,000-square-kilometre site is too small to sustain a reasonable mining operation."[21] Comparison of these net nodule recovery estimates with those of Deepsea Ventures 1971 (Table 1) and the AMC's 1973 figures (Table 2) reveal only modest changes, and thus it would appear that the major reason for the increased area requirements relates to an increase in production factors.

2.4 United Nations Conference on the Law of the Sea

As yet, little substantive discussion of mine-site area requirements has taken place in the Conference. At least two area requirements, however, have been mentioned; the first, contained in a 1970 United States Working Paper, provided for an area of 10,000 km^2, and a 20-year production period which could be renewed for another 20 years. The second was included in the Committee I Chairman's 1976 Introduction to the Single Negotiating Text, where it was indicated that "mining experts" had advised that "...the area needed for a 3 million tons a year operation for 20 years is likely to be about 40,000 sq. kms."[22] Assuming the units are dry metric tons, the net nodule recovery is approximately 1,500 dmt/km^2.

3. Evaluation of Mine-site Area Requirements

3.1 Assessment of Factors

3.1.1 Considerations

Although the worth of a critical evaluation of industry information should not be underestimated, it still remains that virtually all factor estimates come from industry and must be taken more or less 'as is'. In some instances, reference to academic or government data can provide a crude check of industry estimates. The limitations of information from these sources is understandable, as it has been collected with other purposes in mind and thus, for the most part, is inappropriate and/or inadequate. Even where applicable independent data exists, sound interpretation requires background information and expertise that generally does not exist outside the ocean mining industry.

Any attempt to reach a "consensus" from the information published by industry is, at least upon initial review, thwarted by the apparent diversity of figures. To the extent that a consensus evaluation has been developed in this paper, the following background should be borne in mind:

1) Terminology is, in many instances, imprecise; factors are ill-defined and the limits of their application are not noted.
2) What in some instances appears to be a consensus, actually represents a single corporate or individual opinion that has been repeated without identifying the primary source.
3) Real differences of opinion among industry sources legitimately arise from the fact that their operations are different (mine-sites, methodologies, mining systems, products, and production).
4) Some of the information has been provided verbally by, for example, interviews, congressional testimony, and it may represent a very general estimate rather than a succinct evaluation.
5) Industry opinion has evolved and is still evolving as new information is collected and evaluated.
6) Corporate and personal interests may bias the information provided.

3.1.2 Production Factors

3.1.2.1 Recovery

Annual nodule recovery requirements are determined by a company after careful consideration of the capabilities of its mining system, the economics of producing at various scales and its ability to market the resulting product volume and blend. Different approaches towards products and markets, as well as technology, result in different optimum recovery rates. The total annual recovery suggested by the various ocean mining ventures ranges from slightly less than 1 million dmt/year to slightly more than 3 million dmt/year (Table 3). Occasionally, larger annual recoveries are mentioned.[23] Total annual production may, in some instances, be the total production from more than one individual mining unit, again, reflecting differing opinions of system capability, reliability and market availability. The economics of an ocean mining venture will be extremely sensitive to the annual recovery rate. Because a sea-bed mining operation is an integrated operation, it is critical that a resource management regime allow sufficient annual recovery to provide adequate feed to a large-scale processing facility so that advantage may be taken of higher economic efficiencies associated with larger scale facilities. For the purpose of this evaluation, an annual recovery rate of 3 million dmt/year is utilized in all examples unless otherwise noted.

TABLE 3.
Summary of Production Factors Envisaged by Existing Ocean Mining Research and Development Consortia

Ocean Mining Consortia	Total Annual Recovery (x 10^6 dmt/year)	Duration (years)	Cum. Recovery x10^6 dmt	Source
Ocean Mining Associates:				
one initial unit producing at 1 million dmt/year; additional or replacement units in time.	.9 initially; expanding to as much as 2.7	20 to 40 (see text)	36 to 109 (see table)	various, see text, Deepsea Ventures Inc.
Kennecott Consortium:				P.C. Kennecott
one unit	2.7	——	——	Copper Corp., 1975
Ocean Management Inc.:				
two units, each producing approximately 1.5 million dmt/year	3.0	at least 20	60	Shaw[37]
French Association:				
two units, each producing approximately 1.5 million dmt/year*	3.0	25	75	Le Marchand[38]
Continuous Line Bucket Consortium:				
estimates of the recovery capability of a single—2 ship CLB system vary considerably	1.8	——	——	Mero[39]
Amoco Consortium**:				
not specified, have noted only "typical" recovery of 4500 dmt/day per unit				Wenzel, J.G., Testimony before U.S. House Representatives Subcommittee on Oceanography, April 19, 1977

*Uncertain that description in source applies to their envisioned operation.
**At the time of writing there has been no indication that Consortium agreements have been finalized.

3.1.2.2 Duration

Industry recommendations for mine-site area requirements have been based upon durations of 20 and, more frequently, 40 years. It is unfortunate, in view of the fact that this factor has contributed so much to the variance of estimated mine-site size requirements that the justification, particularly for the 40-year duration, has not been better documented. The AMC, one of the consistent users of 40 years duration, simply suggests that "For the most part, industry would like to think in terms of at least a 40-year mine life."[24] In other cases, 40 years, is "...used merely for the purpose of aiding in the mechanical presentation of data."[25] Deepsea Ventures provides some of the few comments of any substance in support of longer duration—"Applying reasonable and prudent approaches to the conservation of equipment, an operation of this sort must be planned for at least a 30-year life and more likely longer."[26] This line of reasoning appears to have originated from their intent to write off, based upon the U.S. Internal Revenue Code, financial commitments on "...pier facilities, building, and main plant facilities..." over a 40-year period which was felt to be "...a reasonable time period based upon the durability of a fair proportion of this equipment."[27] It is economically advantageous to depreciate capital equipment as rapidly as possible for tax purposes. There seem to be no guidelines requiring the depreciation to be spread out over such a long period. Therefore, for the vast majority of capital equipment there appears to be no justification, in terms of providing sufficient duration for a viable operation, for such long mine life.

In view of this, the question of adequate duration might be more reasonably approached from the standpoint of providing at least a fair minimum.

To meet the investment criteria of financing organizations, a mine-site must contain sufficient recoverable reserves to cover projected production volume over the financing life plus a sizeable protective surplus cushion that should be at least 100 percent.[28] Depending upon when debt financing is obtained in relation to a project's schedule, the types of loans obtained and the pay-back period, such reserve requirements could easily amount to a period of the order of 15-20 years at envisioned recovery rates. From the corporate standpoint, the duration should not be so short as to depress an otherwise reasonable and attractive rate of return. Corporate participants of various consortia such as Rio Tinto Zinc and Sedco Inc.,[29] have indicated that rates of return in the order of 15% (D.C.F.) barely meet the minimum requirements for investments in such high risk ventures as deep sea-bed mining. Reflecting a similar feeling, Deepsea Ventures Inc. estimated that the return on investment (D.C.F. after taxes) is marginal for the risks involved and ranges from 5-15% for three-metal recovery and 10-18% for four-metal recovery.[30] Other industry representatives, reluctant to cite projected rates of return have, when questioned, referred to U.S. Department of Interior studies[31] that estimated rates of return (D.C.F.) for a 3 million dmt/year operation producing three metals (nickel, copper and cobalt) at 12.6 to 22.7%. Such rates of return associated with a capital intensive deep sea-bed mining venture will be extremely sensitive to durations of less than 20 years, for which they will decrease significantly.

For the same operations, rates of return are, for practical purposes, insensitive to extensions of duration beyond 20 years, and thus if evaluated strictly by rate of return, there is no justification for requirements of 30 or 40 years. Projects of greater duration, however, could be attractive to industry in the event that the usable life of capital equipment is extremely long and provided that technical obsolescence will not have made the operation of the capital equipment beyond 20 years an uneconomic proposition.

The opinion is then forwarded that for an operation that satisfies industry's rate of return requirements, 20 years duration is adequate to meet the requirements of financing institutions, and would be adequate to allow the maximum practical rate of return for the project. It must be emphasized as well that 20 years is also a minimum, and that a reasonably efficient operation must be allowed sufficient mine-site area to meet a 20-year production requirement.

3.3 Site Factors

The majority of industry's site factor information results from their intensive efforts in the northeast equatorial Pacific. With the exception of the French Association's work in high grade areas in the South Pacific, industry data on other deep sea-bed areas is rather limited and, in most instances, only slightly exceeds that which is available in the published literature. It should also be noted that while regional work has spanned the northeast equatorial Pacific area, budget constraints force a company to focus its more detailed efforts on a very limited number of relatively good prospects. Thus, even industry data has its limitations when it must be generalized to account for possible conditions of all potentially workable sites and not just the best sites. This limitation is offset to some extent as industry does have a better understanding of technological limitations that would, for some factors, bound the conditions characterizing workable sites.

3.4 Mineable Portion

Estimation of the proportion of a mine-site area in which mining would be attempted is made by eliminating both areas that are topographically unfavourable as well as areas that are below cut off abundance and/or grade which can be delineated and avoided. The mineable portion of the site area (m) is then

$$m = 1 - \left(\frac{A_u + A_c}{A_s}\right) \qquad (7) \qquad A$$

where:

A_u = area that is unmineable due to topography, obstances and other bottom conditions[32]

A_c = area below cut off grade and/or abundance that is largely enough to be delineated and avoided.

A_s = total mine-site area.

The approach is adopted from Flipse et al.[33] Although they do not clearly state that the relationship is valid if and only if areas A_c are outside areas A_u ($A_u| A_c \wedge A_u = \phi$), it is implied in several of their basic relationships.

Deepsea Ventures indicates that, based upon preliminary analysis of several areas, 75 to 80% of a mine-site can be considered mineable.[34] Figures of 75 to 85% mineable area are given by both Flipse et al (1973) and Moncrieff and Smale-Adams (1974; for "first generation operations"), while the broadest range is given by Engo (1975) as being 60 to 90% based upon the advice of "mining experts." Within the accessible areas of some mine-sites, Flipse et al (1973), Kaufman (1974) and Siapno (1975) point out the existence of areas containing nodules below cut-off grade and/or abundance. Where such areas are large enough to be delineated and if they are excluded from the mineable portion of the deposit in the mine plan, they should be taken into account. A general value of 10 percent is used by all sources noted above, but it is suggested by Kaufman that "In a typical mine-site, about 10-20 percent of the deposit may contain concentrations too low to be mined,"[35] and that in some candidate mining sites, areas with less than a minimum assay could be as much as 30%. Independence of the two types of areas is not discussed. Although Deep Tow work by the Marine Physical Laboratory at Scripps Institution of Oceanography has provided some of the most accurate maps of detailed bathymetry from the northeast equatorial Pacific, the lack of information on mining system capabilities and mining plans largely precludes the use of this information in determining what constitutes mineable area. An evaluation of two surveyed areas was made by eliminating areas of high slope and potential obstacles and yielded residual areas within the range provided by industry. An approximate range for the mineable portion, excluding both inaccessible areas and these below cut-off grade, would be in the order of .55 to .90 and have an inner, more probable range of .70 to .85.

3.5 Abundance

Discussions of abundance have been fraught with problems of terminology: failure to distinguish wet or dry basis, confusion with "coverage" and "population," and synonymous use of "density," "concentration" and "areal density." All this has caused problems for both the informed and the uninitiated. While these appear to be coming more or less under control, further problems are becoming apparent. Recalling from a previous section that the mineable portion (m) is arrived at after eliminating areas that are topographically unfavourable (A_u) or contain nodule patches below cut-off grade and/or abundance (A_c), it is evident that the sizing equations are valid if and only if the average abundance in the mineable portion (a_m) is used and not the average abundance over the entire mine-site (a_s), which is also a function of the average abundances in unmineable areas (a_u) and in below cut-off grade and/or abundance areas (a_c). Notably, virtually all of the "average abundance" data provided in the literature has been presented as applying to the "mine-site" and forced acceptance of the assumption that $a_m = a_s$, or that a_m was actually meant.

In the event the information actually represents the average abundance for the total mine-site (a_s), then some appreciation of the error introduced can be obtained by rearranging the basic equality,

$$A_s a_s = (A_m a_m) + (A_n a_n)$$

such that:

$$a_m = \frac{a_s - a_n}{m} + a_n \qquad (8)$$

where:

A_s = total mine-site area (km²)

a_s = average nodule abundance in the total mine-site area (dmt/km²)

A_m = mineable area (km²)

a_m = average nodule abundance in mineable areas of a mine-site (dmt/km²)

m = proportion of a mine-site that is mineable

A_N = area not mineable (km^2), note: $A_n = A_u + A_c$
a_N = average nodule abundance in portions that are not mineable (dmt/km^2)
n = proportion of a mine-site that is not mineable

Using equation (8), a_m can be expressed in terms of a_s as a function of m for several relationships of a_n and a_s (Figure 1). As might be expected a_m approaches a_s as m increases and the difference between a_n and a_s decreases. As long as the average abundance outside mineable areas does not vary by more than ± 25% of the average abundance in the total site and the mineable portion is more than 0.7, the error of using a_s for a_m will be less than $\simeq$10% of a_s. Although values for m below 0.7 will apparently be rare, it is not certain that average abundances in the portion of a site that is not mineable will approach the site average. By definition, very low abundances can be expected in areas excluded from the mineable portion due to inadequate abundance, but whether or not this effect of lower a_c is offset will naturally depend upon the relative sizes of A_u and A_c (equation 7) and the average abundance a_u.

FIGURE 1. Graph of Values of Average Abundance in Mineable Areas Expressed in Terms of Average Mine Site Abundance, As a Function of Mineable Proportion.

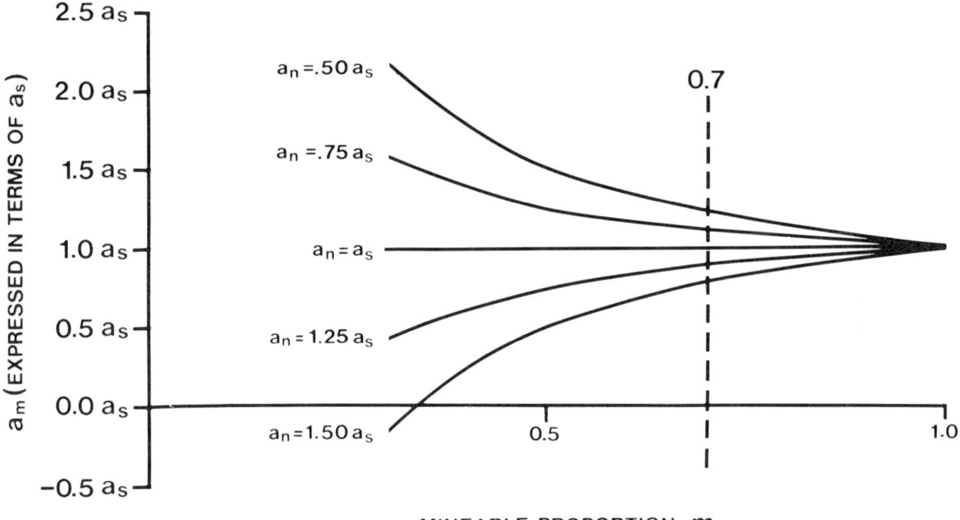

Another area of emerging difficulty is the distinction between cut-off abundance and minimum average abundance. When used in the context of mine-site size requirements it appears that the problem could be resolved by more clearly defining which is being used in association with an abundance figure. A review of industry information rather consistently yields a figure of 6,500 dmt/km^2 as being the "most probable" average abundance in a typical mine-site.[37] The figure has also been adopted by Holser (1976) and Archer (1977) as approximating minimum average abundance.

As to the range of average abundance that might be encountered in a variety of individual sites, industry estimates provide an unsatisfying broad range of possibilities. Drawing on information obtained by Kennecott Copper Corp., Moncrieff and Smale-Adams[38] estimate the approximate limits of abundance "...that might be expected for first generation projects," as 8,000 to 15,000 dmt/km^2 and 10,000 dmt/km^2 as a "medium." The results of exploration by the French Association[39] on the other hand, indicate abundances in excess of 6,500 dmt/km^2 are very rare, and suggest that the "...often advanced..." minimum average abundance of 6,500 dmt/km^2 is too high. Deepsea Ventures estimates that the maximum realistic average abundance that could be found over large areas is 9,800 dmt/km^2. On the low side, apparently referring to cut-off abundance, they suggest that 3,200 dmt/km^2 is the lowest that could "feasibly" be mined with presently envisioned systems.

Putting aside problems of terminology and the differences that could arise if attention were focused on different mine-site regions, there appears to be a real divergence of opinion at basic levels of data interpretation. As it is well beyond the scope of this paper to examine the problem, suffice it to say that given the nature of the resource, varying approaches to sampling methods, sample distribution, interpretive methodology, methods of statistical evaluation and mine planning could easily account for the range of possibilities. Evaluation of other published abundance data derived from "large" area samples (e.g. box corers, and grabs as opposed to piston corers and bottom photographs) tends to support an upper bound of 10,000 dmt/km^2 as suggested by Deepsea Ventures, but allows for some possibility of a limited number of higher abundance sites. This coupled with the data from the French Association and the restriction of Moncrieff and Smale-Adams to consideration of "first generation projects," suggests that 10,000 dmt/km^2, as an average abundance in the mineable portion of a specific mine-site, will only occasionally be exceeded. A range of 6,500 to 10,000 dmt/km^2 is, therefore, assumed.

3.6 System Factors

The combination of dredge and sweep efficiency will determine what portion of the theoretically available nodules can actually be recovered. Only industry sources are cited herein as other estimates would represent guesses, lacking the information from modelling experiments or prototype tests. Better information concerning these factors will probably be made available by industry following the deep water prototype tests of Ocean Mining Associates and Ocean Management Inc. during 1977 and 1978.

3.6.1 Dredge Efficiency

Ranges for dredge efficiency, usually expressed as a percentage, are broad. Some investigators suggest a range of 30% to 70% based upon "...studies and actual experience with dredge baskets and hydraulic equipment..."[40] A range of 40% to 80% is suggested by Moncrieff and Smale-Adams and by Engo. The upper limit is in agreement with Gauthier's estimate for CLB System dredge efficiency.[41] Kaufman suggests that "...equipment design and other decisions predicated upon dredge head pick up efficiencies greater than 60% would be unduly optimistic."[42] Estimates provided by industry represenatives on the DOMES Advisory Panel are in the ranges of 50% to 85%.[43]

The full range of estimates is .30 to .85, but there appears to be an inner, more probable range of .40 to .70.

3.6.2 Sweep Efficiency

A review of statements regarding sweep efficiency suggests general consensus that the minimum will be between 40% and 45%, and the maximum between 65% and 70%. The narrowest range is adopted by Deepsea Ventures, noting that "... (their) studies indicate that with carefully planned dredging procedures and patterns and bottom navigation systems as accurate as is likely to become available, sweep efficiency of 65% is the maximum that can be expected and still maintain required production rates",[44] and that a minimum of 45% might be expected. A similar number is provided by Flipse et al noting that actual values differ somewhat with different mining systems. Moncrieff and Smale-Adams and Engo indicate a range of 40% to 70%. The broadest range, 40% to 75%, has been provided by industry representatives on the DOMES Advisory Panel as being typical values.

The maximum range of .40 to .75 is adopted for use herein.

3.2 Discussion of Area Requirements

The size of a mine-site that would be required in order to provide adequate reserves to meet a specific set of production factors and thus a known cumulative recovery requirement, can be determined by the following relationship: $A_s = \frac{r_a d}{R}$

Because the net nodule recovery, R, is in turn a function of site and system factors which are variable (equation 3), net nodule recovery and thus mine-site area requirements will not be constant, but may take on a range of values corresponding to the possible combinations of site and system factors.

Monte Carlo techniques are used in this paper to combine individual factor probability distributions arrived at in the previous section (Table 4) according to equation (1). These yield the probable distribution of values of R.

TABLE 4
Site and System Factor Distributions For Seabed Mining Operations

Factor	Values	Distribution
mineable portion (m)	.60-.70-.85-.90	trapezoidal
average abundance (am)	6500-10,000	rectangular
dredge efficiency (ed)	.30-.40-.70-.85	trapezoidal
sweep efficiency (es)	.40-.75	rectangular

The resulting probability distribution for (Figure 2) is simply an indication of the chances that R will take on certain ranges of values. For instance, there is a probability of 0.05 that R will exceed 3380 dmt/km^2 and a probability of 0.05 that R will be less than 1090 dmt/km^2. Phrased differently, there is a 90% chance that R will fall between these two values, which form the 90% confidence interval. It must be emphasized that this probability distribution for R is dependent upon the distributions assumed for the site and system factors. In view of the nature of those distributions the probability distributions of R and those derived from the distribution should be taken as approximate and primarily for purposes of reasonable illustration.

FIGURE 2. Probability That the Net Nodule Recovery (R) of a Randomly Selected Operation Will be Greater Than a Certain Value.

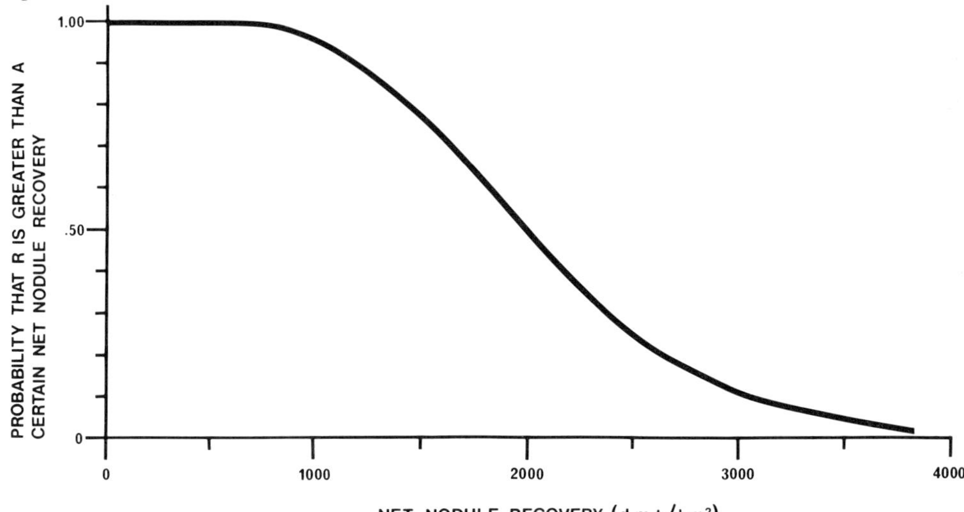

As would be expected, in that similar information was utilized, various industry approximations of net nodule recovery easily fall within the most probable range for R (see Tables 1 and 2).

Using equation (9), the distribution of R is combined with cumulative production requirements for operations recovering 1 million and 3 million dmt/year for 20 years (Figure 3). Considering the 3 million dmt/year operation, the distribution indicates that for any single mine-site selected at random from all possible sites, the chance that the area of that site would be less than 17,800 km^2 or more that 55,000 km^2 is only 10% and there is a 90% chance of it falling between the two values.

FIGURE 3. Probability That the Mine Site Area Requirements (A_S) of a Randomly Selected Mine Site Will be Greater Than a Certain Area.

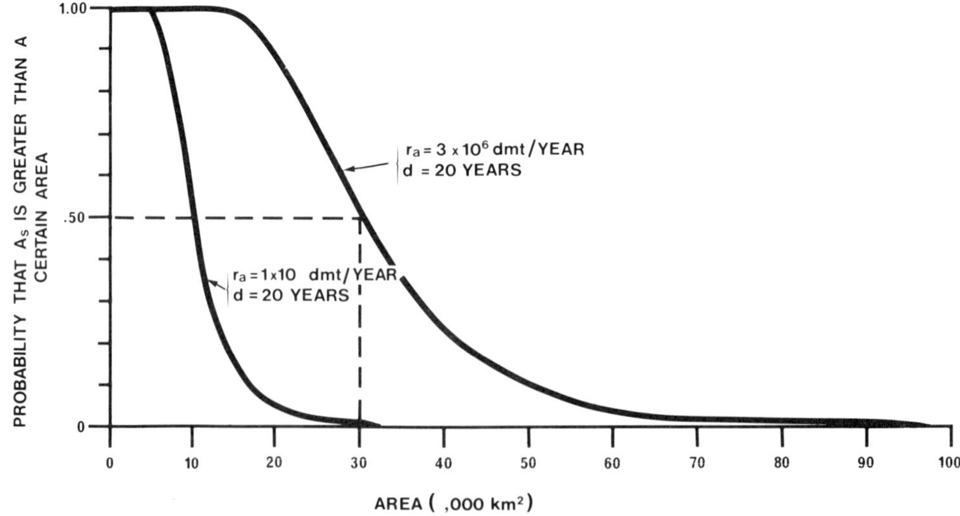

A probability distribution for net mining efficiency, E, has been constructed using the same Monte Carlo methods and the appropriate factors (Table 4) as required by equation (6) (Figure 4).[45] The distribution indicates that the 90% confidence interval for E is 0.15-0.40. Comparison with a number of estimates that are available for net mining efficiency indicates good agreement. Holser, having the benefit of proprietary input, distinguishes between the net mining efficiency of first, second and third generation operations, and suggests 20%, 35%, and 50% respectively. Thiry et al believe that first generation mining equipment will probably be less than 20% efficient, Amann assumes 20%, and Archer suggests that first generation mining equipment is likely to be between 10% and 50% efficient, but perhaps more likely between 20% and 25%. Similarily, based upon the advice of "mining experts," Engo suggests a maximum range of 10% to 50% and a more probable range of 20% to 30%. By directly combining individual estimates of site and system factors for first generation operations Moncrieff and Smale-Adams utilize a low of 12%, a medium of 26% and a high of 48%.

FIGURE 4. Probability That the Net Mining Efficiency (E) of a Randomly Selected Mining Operation Will be Greater Than a Certain Value.

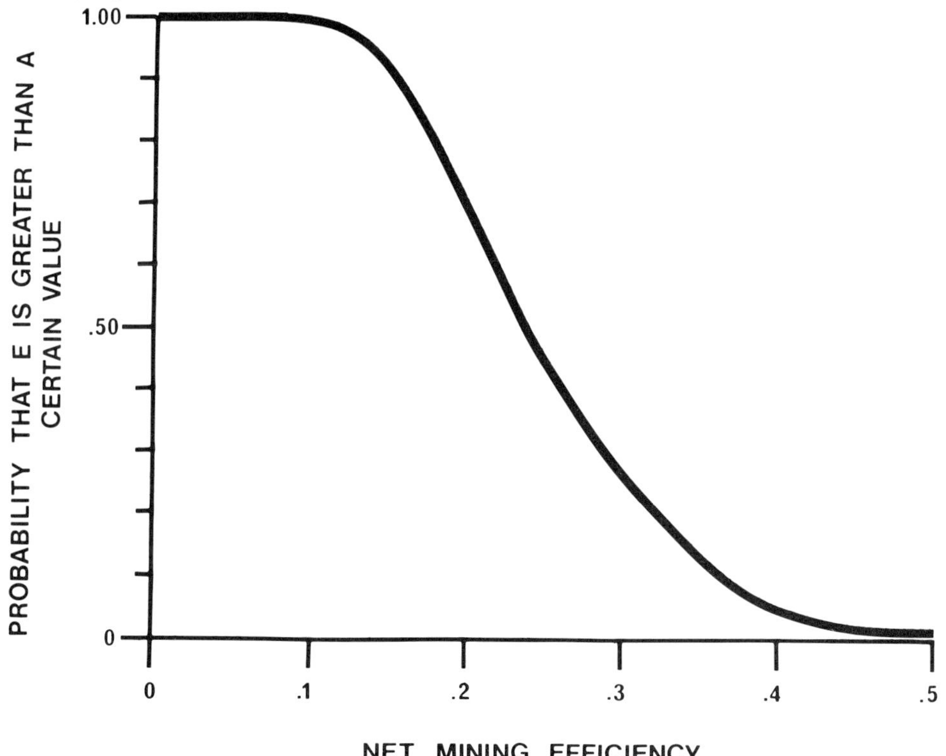

Even if we now consider the case of a specific sea floor area and a particular mining system, net nodule recovery will still not be known exactly as there remains, even for a specific case, some degree of uncertainty associated with each individual site and system factor.

At the time a company applies for a development license, a considerable amount of exploration will have been conducted within the region that will constitute the mine-site. While the information gathered will provide a company with a reasonably good idea as to how much of the site is potentially mineable and what might be the average abundance in mineable areas, the nature and cost of such exploration will not permit the developer to state with certainty, single precise values for these factors. In the case of abundance, measurement errors associated with even the most reliable of techniques, coupled with the natural variability of the "true" abundance measurements, results in a range of possible values within which the correct average abundance probably falls.

Similarly, for the case of system factors, prototype mining tests will provide a reasonably good indication of the dredge and sweep efficiencies that could be expected for the particular mining system to be utilized. The results, however, will again provide only a range within which dredge and sweep efficiency will probably fall.

The site and system factors listed for case A in Table 5, for instance, might represent the best estimates available at the time a development licence is applied for. If the operation requires sufficient area to ensure 20 years of recovery at a rate of 3 million dmt/year, the area requirements (Figure 5) might be anywhere between 39,200 km^2 and 58,600 km^2 (90% confidence interval). Because of the valid uncertainty associated with factor values, any narrowing of the limits increases the probability that the correct area requirement will fall outside the narrowed range.

TABLE 5.
Site and System Factor Distribution for Specific Seabed Mining Operations

	Case A	Case B
mineable portion (m)	.75-.85 (r)	.75-.85 (r)
average abundance (a_m)	6,500-8,000-9,500 (t)	6,500-8,000-9,500 (t)
dredge efficiency (e_d)	.35-.45 (r)	.45-.55 (r)
sweep efficiency (e_s)	.45-.55 (r)	.55-.65 (r)

( ): distribution
r : rectangular
t : triangular

Mine-sites and mining systems will differ from one another to some extent and thus will have different ranges for the area required to meet their individual production requirements. For example, assuming a more efficient mining system is operating on the same site as in the previous example (Table 5 case B), the distribution in Figure 5 indicates that the area requirements should be not more than 38,800 km^2 and not less than 26,400 km^2.

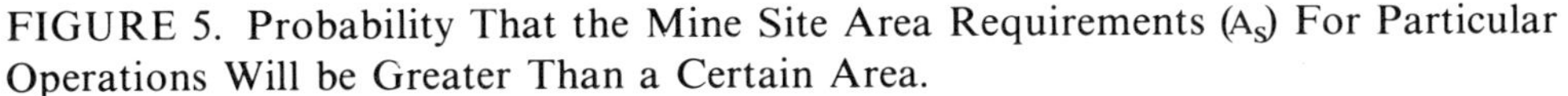

FIGURE 5. Probability That the Mine Site Area Requirements (A_S) For Particular Operations Will be Greater Than a Certain Area.

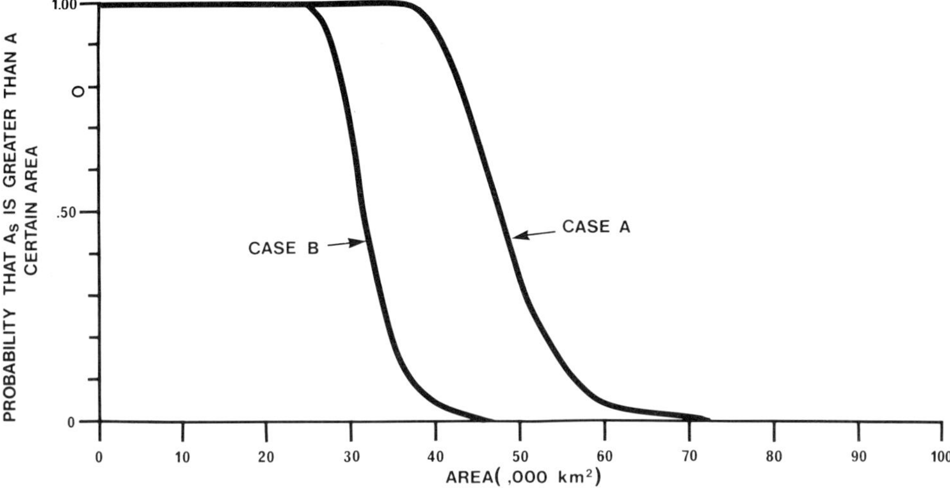

4. Methods of Defining Mine-site Area Requirements

4.1 Alternatives and Evaluation Criteria

The United Nations Secretary-General's 1974 "Report on Economic Implications of Sea-bed Development in the International Area"[47] suggests three possible alternative approaches to the determination of mine-site area requirements: 1) a standard size for all mine-sites regardless of locality or proposed production capacity, 2) variable size determined by the requirements of the sea-bed miner and 3) variable sizes predetermined by a master grid. Other options include a limit on the cumulative recovery from an operation and, as suggested by La Que and Inco Ltd.[47], a system that would fix the size of an area but provide for the renegotiation of the location should the site prove unproductive or otherwise unattractive.

It must be emphasized that the method for determining mine-site area requirements is but one of many elements that will comprise a complete resource management system. Such items as limits on the number of sites held by any consortium or State, environmental regulations, compatibility with the concepts of "banking" mine-sites and a ceiling on total annual production from all sea-bed operations, work requirements, and fees, would, depending upon the regime adopted, have to be considered. Attention herein, however, focuses specifically on the merits and possible problems of defining mine-site area requirements by the above noted methods. The basis of evaluation criteria and considerations is as follows:

(1) *sufficient area* should be provided in order to ensure that minimum requirements for annual recovery and duration are met. In view of the sensitivity of project economics to annual recovery and durations of less than 20 years, uncertainty of meeting minimum requirements could inhibit if not stifle the development of the resource. Based upon the previous discussion, recovery rate and duration utilized

herein will be any reasonable annual recovery that would be required in order to take advantage of large scale economics and a minimum of 20 years duration.

(2) *Oligopolistic situations* should be discouraged by maximizing, to the extent practical, the number of mine-sites available for exploitation. This may be particularly important during the early stages of sea-bed development, in view of the possibility that the number of sea-bed sites capable of supporting economically competitive operations may be somewhat more limited than previously envisioned.

Although the deep sea-bed nodule resource is probably large, there exists some possibility that at least the immediately exploitable resource is limited and thus calculation of site requirements based upon excessively large cumulative production could result in either corporations, States or groups of States holding a disproportionately large share of the immediately exploitable resource.[49] As the quantity of resource that might be recovered over the long term is probably considerable, preventing an oligopolistic situation by promoting diversification of holdings at an early state of resource development may be desirable in order to discourage the possibility that only a very small number of companies and States would have the opportunity to improve and refine the techniques that might be necessary to economically exploit the larger quantity of somewhat less favourable resource.

A comparative evaluation of the effect that various sizing methods might have on mine-site availability is difficult due to the uncertainties involved. In order to provide some relative indication of the effect, it will be necessary to assume a single value that corresponds to the area in the world ocean where nodules of adequate grade and abundance tend to co-exist. Due in large part to differences in defining and applying cut-off criteria, there is virtually no value that meets with common acceptance. Solely for purposes of discussion, Archer's 1977 estimate of 4.25 million km^2 is adopted.[50] Table 6, which lists the frequency and number of individual areas that would be capable of yielding 20 years of nodule supply, has been constructed in order to facilitate the comparison of mine-site availability. It assumes that there are 4.25 million km^2 of potentially exploitable sea-bed area and that the probability distribution in Figure 3 is the actual distribution of area sizes with approximately 20 years of recoverable resources. Under these circumstances, there would exist a maximum of approximately 135 viable areas with 20 year reserves of recoverable nodules.

TABLE 6.
Frequency and Number of Individual Areas of Various Sizes That Contain 20 Years of Recoverable Reserves*

Size of Area (km^2) Range	Class Mark	Frequency	Cumulative Frequency	Number	Cumulative Number
10-20,000	15,000	13	13	17.32	17.32
20-30,000	25,000	36	49	47.96	65.28
30-40,000	35,000	28	77	37.30	102.58
40-50,000	45,000	15	92	19.98	122.56
50-60,000	55,000	8	100	10.66	133.22

*Assuming the distribution of probable sizes in Figure 3 and a potentially exploitable seabed area of 4.25 million km^2.

Traditionally the "number of mine-sites" has been derived by using a typical cumulative recovery to convert recoverable resource into the number of mine-sites. While the method is expedient, it ignores the fact that mine-sites will likely be defined in terms of area and probably yield cumulative production significantly different from that which has been adopted as a typical value.

(3) *conservation* of the resources should be promoted so as to maximize the quantity of recoverable nodules. Any scheme should be designed such that industry is encouraged not only to operate at a reasonable minimum efficiency but, in fact to increase efficiency. Traditionally, this has been accomplished by providing for the unlimited renewal of a mine-site subject to efficient recovery from the area.

(4) *administrative aspects* such as requirements for highly skilled personnel and the acquisition of an independent data base will affect both the cost and effectiveness of the options.

(5) *acceptability* of the method by both the organization responsible for the management of the resource and the entity whose skills and investment will be required to develop the resource will be necessary. A critical aspect of acceptability from the standpoint of an operation relates to the extent to which the resource manager has powers to fix or adjust area requirements.

4.2 Discussion of Alternatives

4.2.1 Fixed Area Sites

The fixed or standard mine-site area is quite common. The United States proposed its incorporation in the United Nations Convention on the International Sea-bed Area[51] and it has become a traditional aspect of United States Deep Sea-bed Mineral Resource bills.

Using, for illustrative purposes, the proposal of the American Mining Congress for sea-bed mine-sites of 30,000 km^2 and referring to Figure 3, it is evident that approximately 50% of the potentially exploitable sea-bed sites must be larger than 30,000 km^2 and thus if limited to 30,000 km^2 would not contain sufficient recoverable resource to meet the minimum production requirements of a 3 million dmt/year, 20-year operation. Interestingly, an operation achieving the AMC's "anticipated efficiency" of 1600 dmt/km^2 could produce for only 16 years at 3 million dmt/year from a 30,000 km^2 site, and in fact, only operations having net mining recoveries in excess of 2000 dmt/km^2 would be able to produce from such a site for 20 years or longer.

If the standard area was increased to say 55,000 km^2, 95% of the sites could meet production requirements. Use of a standard site area of 55,000 km^2 and assuming the distribution for net nodule recovery in Figure 2 suggests that roughly 50% of the operations could produce for more than 36 years and 25% could produce for more than 45 years. Although the numbers are illustrative, the example points out that establishing any fixed mine-site size that will provide for minimum production requirements for the vast majority of possible sea-bed operations will result in the more favourable mine-sites and operations having longer mine lives.

Turning to the question of mine-site availability, it will, as previously noted, be assumed that there are approximately 135 mine-sites of various sizes (Table 6), all capable of producing 3 million dmt/year for approximately 20 years.

If a standard 30,000 km^2 were utlized, roughly 50% of the areas (by number) that contain sufficient reserves for 20 years duration are larger than 30,000 km^2 and could not be exploited. Thus, only approximately 65 adequate size areas would be available. The 65 areas, converted to total area and divided by 30,000 km^2 (as any 30,000 km^2 portion of the area would support at least 20 years of production), indicates roughly 50 (30,000 km^2) areas exist that would contain reserves for 20 years or more. Approximately two-thirds of the potentially exploitable seafloor area could not be exploited if mine-sites were limited to 30,000 km^2.

On the other hand, there are very few areas capable of supporting a 20-year operation that are larger than 55,000 km^2, thus virtually the whole of the 4.25 million km^2 could be broken up into 55,000 km^2 areas, each with at least 20 years of resources. Therefore, the number of viable 55,000 km^2 sites would approach approximately 80.[52]

The rather unsettling indication that there is greater availability of larger sites capable of supporting a 20-year operation stems from the conclusion that a significant portion of potentially exploitable seafloor (roughly two-thirds, in this example), mined with presently envisioned systems, would not contain 20 years of recoverable reserves if mine-sites were limited to 30,000 km^2, whereas virtually any 55,000 km^2 potentially exploitable sea-bed area, mined with envisioned systems, would probably contain at least 20 years of recoverable reserves.

Establishing a standard mine-site size on the basis that it should provide adequate reserves under all reasonable recovery conditions will allow the possibility that a significant portion of the operations will be of very long duration. It may well be the case, however, that the use of standard areas derived in this fashion increases the quantity of exploitable resource and the number of viable mine-sites as compared to smaller fixed areas that decrease the possibility of long mine lives, and in many instances, will probably not contain sufficient reserves to be exploited.

As previously noted, consideration must also be given to the possibility that the portion of the total resource that is exploitable in the near future may be rather limited and/or is concentrated in relatively small areas. Under such circumstances, the use of large "standard" site areas would decrease the number of readily exploitable sites and enhance the possibility that that portion of the resource could be controlled by a small number of companies or States. It is unknown whether auxiliary provisions limiting the number of early mine-sites held by an entity (company, consortium, State, etc.) could or would be needed to overcome this potential difficulty.

Providing a fixed mine-site area that can be exploited so long as the operator desires would provide considerable incentive for a company to increase mining efficiency, so as to increase the longevity of the operation. If a specific "standard" mine-site of 55,000 km^2 was in fact 80% mineable (m = .80) and had an average abundance in mineable areas of 8,000 dmt/km^2, an increase in dredge efficiency from .40 to .50, and in sweep efficiency from .50 to .60 would increase the productive life of a 3 million dmt/year operation from 23 to 35 years. Administrative responsibilities and costs associated with the use of standard area development licences would be minimal.

So long as the area were sufficient to assure the minimum cumulative recovery requirement, industry would probably find the system acceptable. The resource management agency, on the other hand, would have to evaluate the acceptability of large sites that would allow some operations very long durations and might result in a small number of operators controlling the most readily exploitable portion of the resource, against the loss of exploitable resource contained in standard sites having insufficient reserves.

4.2.2 Cumulative Production Limit

Although it may not have been specifically suggested, proposals occasionally have been made that would, assuming a specific annual recovery, in effect, set a limit on cumulative production from a mine-site by a particular operator, thereby establishing a maximum duration. One version of a proposed United States Deep Sea-bed Hard Minerals Act stated that an exploitation "... license shall remain in force as long as commercial recovery from the block continues or until the end of the 20th year subsequent to beginning commercial recovery, whichever comes first."[53] More commonly, such time limits are more generous and are established by limiting the number of renewals.[54]

Considered herein is the case where a standard mine-site area is given on the basis of a specific annual recovery and limited duration. Obviously, any mechanism that allowed an operator to produce at his favoured rate for 20 years or longer would need minimum duration criteria so long as the mine-site area were adequate. If it is assumed that companies operating under such a system attempted to maintain and improve both dredge and sweep efficiencies, the method would conceivably result in maximum availability of mine-sites. As earlier noted, computations of numbers of mine-sites have, in fact, assumed for convenience that this is the case, as computations involve the direct conversion of recoverable resource to mine-sites based upon a specified cumulative recovery. This assumption regarding efficiency is fallacious and, in fact, a major disadvantage of the method stems from the fact that an operator, realizing he is limited to a given production, has little incentive to increase his mining efficiency and instead would have excellent economic reasons to mine high grade material within the assigned area and/or turn to less efficient, inexpensive mining methods.

Returning to the previous example, mining an area having an average abundance in mineable area of 8,000 dmt/km^2 and that is 80% mineable, at dredge and sweep efficiencies of .40 and .50 respectively, would require 46,900 km^2. If the dredge and sweep efficiency could have been improved to average .50 and .60 respectively, only 31,300 km^2 would have been mined out in 20 years. As remining may not be feasible, the potential loss of recoverable resource from that single site would be 30 million dmt—50% of the actual cumulative recovery if calculated on the basis of 3 million dmt/year. The net result of inefficient operations would be to decrease both the amount of recoverable resource and the number of mine-sites. The potential loss of recoverable resource and mine-site availability would probably be considerably less than if the operator benefited directly from the results of his efforts to achieve the highest possible recovery efficiencies.

Assuming that a minimum acceptable production efficiency had to be employed, personnel with technological expertise would be required in order to establish minimum limits and ensure that they were met. From any standpoint, the mechanism provides little that would be attractive to either the operator or managing agency.

4.2.3 Redefineable Fixed Areas

Both LaQue's original proposal and the modified version suggested by The International Nickel Company, Ltd., now Inco Ltd., provide for an exclusive development license to a relatively small area which "if the licensee found the mining location defined by the original starting point unproductive and otherwise unattractive, he would have the right to initiate negotiations for redefinition of the area of exclusive access."[55] The combination of provisions which allows renegotiation of the location of a fixed area, while at the same time indicating that "the number of licenses that could be held by a single licensee would be subject to limits set by the licensing authority" imposes uncertainties that prohibit meaningful evaluation.

Given 2,500 km^2 block sizes, as suggested by Inco, that could be relocated without specifying either minimum time period that a block must be held or the maximum number of blocks that could be held cumulatively, the method would seem to virtually promote high-grading and inefficient recovery. In this regard, it is interesting to note that the LaQue proposal included provisions which, to some extent, offset some of these problems by requiring the operator to produce a minimum tonnage over a given period of time and specified a larger area (approximately 8,000 km^2). Acceptability of this concept as proposed, to either the management agency or to an operator, would be questionable. On one hand, it is not certain that minimum production requirements are provided for and, on the other hand, the potential for unacceptable conservation practices is very apparent.

4.2.4 Predetermined Variable Sizes

The use of variable sizes, predetermined by a master grid, is a difficult method to evaluate. It would seem that to have any merit whatsoever, the grid would have to have been designed such that the variable areas were, to some extent, a function of the site factors found in the grid area. Further complexity is provided by uncertainty as to how variations in cumulative production requirements and system factors would be treated.

Lacking more specific information on the intent of this method, evaluation is impossible. At best it appears to be a complex concept that would require a considerable amount of expensive data collection that would provide few, if any, advantages from the standpoint of any of the evaluation criteria.

4.2.5 Variable Mine-site Area

Providing an operator with a mine-site area tailored to the specific site and system conditions expected has been suggested in United Nations Law of the Sea Conference negotiating documents. The Informal Single Negotiating Text,[56] for instance, provided that: "Areas for exploitation shall be calculated to satisfy the production requirements agreed between the Authority and the Contractor over the term of the contract taking into account the state of the art of technology then available for ocean mining and the relevant physical characteristics of the area. Areas shall neither be smaller nor larger than are necessary to satisfy this objective."

As there is virtually no disagreement with the idea that an operator should be provided with an area adequate to yield the necessary cumulative recovery, a mine-site area determined by this method might be calculated so that there would be a high probability that the site contains at least 20 years of recoverable reserves. Because site and system factors will be to some extent uncertain, calculation of the area requirements that yields only a 5% chance that a larger area could be required must, by definition, also mean there is a 95% chance that that much area will not be required. Using Case A (Table 5, Figure 5) as an example, an area of 59,600 km^2 would have to be awarded in order to minimize the chances (e.g. 5%) that the site would contain insufficient reserves. By doing so, it must be accepted that there is a 50% chance that an area of only 47,400 km^2 would be required and a 10% chance that as little as 40,600 km^2 would be actually needed. In the event that only 47,400 km^2 were required, the operator, having a site of 59,600 km^2 could theoretically produce for a total of 25 years, assuming no improvement in system factors.

Theoretically, the method would tend to maximize the number of mine-sites available for exploitation by meeting in each instance the minimum cumulative production requirements. Again, there would exist, according to the example, approximately 135 mine-sites containing sufficient recoverable reserves (Table 6). However, awarding an area that is certain (95% confidence level) to yield at least 20 years of nodules will result in the area probably being larger than is actually necessary and thus there will be fewer than 135 mine-sites available under this approach.

An estimate of minimum site availability for this approach can be made by assuming the area awarded would be at most 50 percent larger than the actual area required. This is based upon areas of sites which are given at a .05 probability that a larger area is required, as compared to the smaller area is actually correct, the maximum potential "over award" is the difference between the two which, after examining several examples (see case A and B for instance; Figure 5) may be up to 50% more than the small area. If all areas awarded are assumed to contain 50 percent more than actually required, a new mine-site area distribution can be constructed and the number of available sites calculated. If there is no maximum limit to an award area, the minimum site availability would approach 80. If the maximum development licence were set at 60,000 km^2, the minimum site availability is slightly under 90. Thus, comparatively, this method of establishing area requirements should provide between 80 and 135 viable sites. Assuming a mine-site area were awarded with no arbitrary maximum duration, the company, as in the case of a fixed mine-site area, would have considerable incentive to increase mining efficiencies in order to prolong the life of the operation.

Determining the individual factor distributions used to compute area requirements would impose upon a management agency administrative responsibility not common to other methods. Assuming that an applicant provided the agency with its estimates for the probable distribution of site and system factors at the time a development licence was applied for, the agency would require expert personnel in order to review the applicant's estimates. Further, because of the very high cost of data acquisition, the agency would probably depend upon the applicant's data and would need virtually total access to any and all information used to derive factor distributions. Obviously, the staff and data evaluation capabilities would lead to higher administrative costs than would be associated with the other options. Although it will not be considered in the present paper, establishing a fair distribution for dredge and sweep efficiencies may be difficult—particularly for the earliest sea-bed mining operations. The possibility gives rise to a host of potential problems and solutions that remain to be evaluated.

The basic approach should be generally acceptable to the operator, as he is virtually guaranteed 20 years of recoverable reserves and allowed to benefit to the fullest extent from improvements he brings about in dredge and sweep efficiency. From the management agency's point of view, the method would also seem satisfactory as it tends to maximize the potential number of mine-sites, thus promoting competition and diversity, while at the same time encouraging the conservation of the resource by allowing benefits of increased efficiency to accrue to the operator. There are, however, increased administrative functions and costs associated with the scheme. Also, potentially significant problems to the agency and/or operator may arise as a particular result of uncertainty associated with dredge and sweep efficiencies.

5. Summary and Conclusions

A sea-bed mine-site must contain sufficient reserves to meet minimum economic production requirements. Given a sea-floor area containing, as a whole, nodules of adequate average grade and abundance, a portion of that area will not be mined due to topographic obstructions, adverse bottom conditions and possibly the existence of patches in which the nodules are below cutoff grade and/or abundance. Considering the part of the mine-site in which mining would actually be attempted, the proportion of the nodules that could be recovered from that part will further be dependent upon efficiencies of the mining system used. Calculating the area requirements for a mine-site must take into consideration these factors as well as the uncertainty associated with their estimation.

There are significant differences between various estimated mine-site size requirements. Such differences arise naturally from the inherent variability of the resource, mine-site characteristics, and the uncertainty associated with estimating and measuring site factors. Variation also stems from differences in envisioned annual rates of recovery and characteristics of individual mining systems. Occasional changes in the estimates made by individual groups can, in some cases, be attributed to a re-assessment of site and system factors in light of new information resulting from exploration programmes and prototype mining tests. A major factor contributing to variation and change has been the doubling of the duration of recovery used to calculate mine-site area requirements from 20 to 40 years.

In large part, industry estimates of the factors that determine mine-site requirements must be relied upon in view of the paucity of applicable data from other sources and the lack of background information with which to evaluate what data are available. Envisioned annual recovery rates vary; large production capacities are favoured in order to take advantage of economies of scale. Technological capabilities and marketing considerations, in some cases, may temper the magnitude of the operation.

Although some potential sea-bed mining organizations have indicated that a 40-year mine life is desirable, no justification is apparent for using durations longer than 20 years in calculating basic mine-site size requirements.

There appears to be a reasonable agreement on the possible range of values of the proportion of a given mine-site in which mining would actually be attempted. Estimates of nodule abundance, on the other hand, vary considerably and are difficult to evaluate because of imprecise definition and failure to specify restrictions and qualifications of individual estimates.

Industry estimates of mining system recovery capabilities, having been developed from engineering studies and prototype tests, differ and will remain tentative until either extensive prototype mining operations are complete or actual mining experience is gained.

Uncertainty in judgments of pertiment factors precludes the possibility of specifying any single mine-site size that will approach the minimum requirements of all operators. Estimates based on published information indicate that some sites with sufficient reserves for a 20-year mine life producing at 3 million dmt/year might be as small as 18,000 km^2 or as large as 55,000km^2 (90% confidence interval). Similarly, even when estimating the area requirements for a specific mine-site and mining operation, there remains an element of uncertainty that again precludes the possibility of specifying exactly the mine-site area requirements that would be needed in order to meet production requirements.

Several methods for establishing mine-site size have been examined and compared on the basis that they provide sufficient reserves, promote resource and mine-site availability, and encourage conservation. Also taken into account are administrative requirements and the acceptability of the options to the developer and the managing agency. The review indicates that a large standard mine-site size or individually calculated mine-site sizes are the most realistic alternatives.

The large standard size has the advantage of simplicity and minimal administrative requirements, but might diminish the near term availability of viable mine-sites and increase the possibility that a relatively small number of companies and/or States could control the most readily exploitable portion of the resource.

Appropriate calculation of each individual mine-site area for each sea-bed mining operation, while tending to make more mine-sites available, would place a considerable administrative burden on the managing agency. Moreover, it remains to be seen whether or not the factors used to calculate area requirements can be determined in a satisfactory manner.

Notes and References

[1]United Nations, Third Conference on the Law of the Sea, 'Note by the Chairman of the First Committee in connection with the Informal Single Negotiating Text," Part I, Contained in Document A/CONF.62/WP.8/Part I, A/CONF.62/C.1/L.16, 5 September 1975.

[2]Flipse, J.E., M.A. Dubs and R.J. Greenwald, 'Pre-production Manganese Nodules Activities and Requirements', *Mineral Resources of the Deep Sea-bed,* Hearings before the Subcommittee on Minerals, Materials and Fuels of the U.S. Senate Committee on Interior and Insular Affairs, 15 March 1973. p. 602-700.

[3]Kaufman, R., 'The Selection and Sizing of Tracts Comprising a Manganese Nodule Ore Body', *1974 Offshore Technology Conference, Preprints.* Vol. II. Paper OTC 2059, p. 283-293.

[4]Siapno, W.D., 'Exploration Technology and Ocean Mining Parameters', *American Mining Congress Convention,* San Francisco, 28 September-1 October 1975, reprint 24 p.

[5]Brooks, D.B. 'Deep Sea Manganese Nodules: from Scientific Phenomenon to World Resource', *Proceedings of the Second Annual Conference,* The Law of the Sea Institute, University of Rhode Island, 26-29 June 1967, p. 32-41.

[6]Christy, F.T. Jr., 'A Social Scientist Writes on Economic Criteria for Rules Governing Exploitation of Deep Sea Minerals', *International Lawyer,* Vol. 2, No. 2, 1968, p. 224-242.

[7]All values have been converted to metric equivalents: 1 metric ton = 2204.6 lbs; 1 km^2 = 0.3861 mi^2; 1 kg/m^2 = 0.205 lbs/ft^2; dry nodule weight = 0.666 wet nodule weight. [2].

[8]Mero, J.L., Alternatives for Mineral Exploitation', *Proceedings of the Second Annual Conference.* The Law of the Sea Institute, University of Rhode Island, 26-29 June 1967, p. 94-97.

Mero, J.L., 'A Legal Regime for Deep Sea Mining', *San Diego Law Review,* Vol. 7, 1970, p. 488-503.

Mero, J.L., *The Mineral Resources of the Sea, Elsevier Publishing Company, Amsterdam, 1965, 313 p.*

[9]*Menard, H.W., Marine Geology of the Pacific,* McGraw Hill, New York, 1964, 371 p.

Moore, T.C. and G.R. Heath, 'Manganese Nodules, Topography and Thickness of Quaternary Sediments in the Central Pacific', *Nature,* 1966, 212: 983-985.

[10]Walthier, T.N., 'Remarks on the Mining of Deep Ocean Mineral Deposits', *Proceedings of the Second Annual Conference,* The Law of the Sea Institute, University of Rhode Island, 26-29 June 1967, p. 98-99.

[11]It has now formed a consortium, Ocean Mining Associates, including U.S. Steel, Union Miniere of Belgium and the Sun Oil Corp. (See Chapter IV C).

[12]Rintoul, B., 'Deepsea Ventures Planning Ocean Mining Test Next Summer', *Offshore* July 1968, p. 41.

Flipse, J.E., 'An Engineering Approach to Ocean Mining', *1969 Offshore Technology Conference, Preprints,* Paper OTC 1035, P. I-317-332.

Flipse, J.E., 'Developments in Ocean Exploration and Mining', *American Mining Congress Convention,* San Francisco, 19-22 October 1969, 19 p.

Rothstein, A.J., 'Deep Ocean Nodule Mining', *Underwater Science and Technology Journal,* September 1970, p. 133-137.

[13]Caldwell, A.B., 'Deepsea Ventures Readying its Attack on Pacific Nodules', *Society of Mining Engineers,* October 1971, p. 54-55.

[14]Deepsea Ventures, Inc., *Notice of Discovery and Claim of Exclusive Mining Rights and Request for Diplomatic Protection of Investment,* filed November 1974 with the U.S. Secretary of State, U.S. Department of State, Washington, D.C.

[15]Greenwald, R.J., 'Problems of Legal Security of the World Hard Minerals Industry in the International Ocean', *The Natural Resource Quarterly,* 1971, Vol. 4, No. 3, p. 639-645.

[16]J.E. Flipse, President, Deepsea Ventures Inc., testimony before U.S. House of Representatives Subcommittee on Oceanography, May 12, 1972; also Metalworking News, December 11, 1972.

[17]J.E. Flipse, President, Deepsea Ventures Inc., testimony before U.S. House of Representatives Subcommittee on Oceanography, March 28, 1973.

[18]Correspondence: T.S. Ary, Chairman, American Mining Congress Committee on Undersea Mineral Resources to Senator L. Metcalf, Chairman, U.S. Senate Subcommittee on Minerals, Materials and Fuels, June 7, 1973.

[19]Correspondence, T.S. Ary, Chairman, AMC Committee on Undersea Mineral Resources, to Senator L. Metcalf, Chairman, U.S. Senate Subcommittee on Minerals, Materials and Fuels, June 7, 1973. "Block", as used in United States Deep Seabed Hard Minerals Resource bills, is apparently synonymous with "mine-site" as used herein.

[20]Correspondence: M.A. Dubs, Chairman, AMC Committee on Undersea Mineral Resources, to Congressman J.M. Murphy, Chairman, U.S. House of Representatives Subcommittee on Oceanography; January 26, 1976.

[21]M.A. Dubs, Testimony on behalf of the American Mining Congress before the U.S. House of Representatives Subcommittee on Oceanography, March 8, 1976.

[22]A/CONF.62/C.1/L.16, *op. cit.*

[23]Amann, H.M. 'Definition of an Ocean Mining Site', *1975 Offshore Technology Conference, Proceedings,* Vol. I, Paper OTC 2238, p. 891-908.

[24]Correspondence: T.S. Ary, Chairman, AMC Committee on Undersea Mineral Resources, to Senator L. Metcalf, Chairman, United States Senate Subcommittee on Minerals, Materials and Fuels, June 7, 1973.

[25]Flipse, et al, 1973, *op cit.*

[26]Kaufman (1974), *op cit.* p. 292.

[27]J.E. Flipse, President, Deepsea Ventures Inc., Testimony before the U.S. House of Representatives Subcommittee on Oceanography, March 1, 1973.

[28]T.C. Houseman, Vice-President, Chase Manhattan Bank, Testimony before the U.S. House of Representatives Subcommittee on Oceanography; March 17, 1977.

[29]B.G. Clements, President SEDCO, Inc., testimony before the U.S. House of Representatives Subcommittee on Oceanography, April 19, 1977.

[30]Kaufman, R. 'Ocean Nodule Mining—Progress and Problems' in McLeod C.R. and D.W. Pasho, Papers presented at the Seventh Underwater Mining Institute, Madison, Wisconsin, 28-29 October 1976. Unpublished Manuscript, Canada, Department of Energy, Mines and Resources.

[31]Wright, R.L., *Ocean Mining: An Economic Evaluation,* U.S. Ocean Mining Administration, U.S. Department of the Interior, May 1976, 18 p.

[32]Precise definition of A_u will have to indicate whether or not the width of a buffer zone between a block limit and an obstacle is included—such a zone is necessary to allow for errors in navigation and control, and thus may vary as a function of both. The present inclination is to include the "buffer zone" in the definition of sweep efficiency.

[33]Flipse et al. (1973), *op cit.*

[34]Kaufman (1974) and Siapno (1975).

[35]Kaufman (1974), page 290.

[36]Moncrieff, A.G. and K.B. Smale-Adams, 'The Economics of First Generation Manganese Nodule Operations', *Mining Congress Journal,* December 1974, p. 46-50.

[37]Anon., 'Definition of Mining System Parameters of Importance to an Environmental Study of Deep Ocean Mining', U.S. Department of commerce, NOAA/DOMES Advisory Panel Meeting, 24 February 1976, 4 p.

[38]Moncrieff and Smale-Adams (1974), p. 46.

[39]Bastien-Thiry, H.B., J.P. Lenoble and P. Rogel, 'French Exploration Seeks to Define Mineable Nodule Tonnages on Pacific Floor', *Engineering and Mining Journal,* July 1977, p. 86-87, 171.

[40]Flipse *et al.* (1973) and Siapno (1975).

[41]Gauthier, M., 'Mining on the Continuous Line Bucket (CLB) System', Paper presented at the European Economic Community Seminar for ACP experts to UNCLOS on the Exploitation of the Deep-Seabed, Brussels, February 1977, 6 p.

[42]Kaufnam, (1974), *op cit.,* p. 287.

[43]U.S. Department of Commerce, *Deep Ocean Mining Environmental Study,* 1976.

[44]Kaufman, (1974), *op cit.* p. 287.

[45]e_d and e_s assumed independent for purposes of illustration.

[46]United Nations, Third Conference on the Law of the Sea, *Economic Implications of Sea-bed Mineral Development in the International Area,* Report of the Secretary-General, A/CONF.62/25, 22 May 1974, p. 80.

[47]LaQue, F.L., 'A Preferred Approach to International Regulation of the Exploitation of Deep Ocean Manganese Nodules', E.M. Borgese (Ed.), 1972, Pacem in Maribus, III, 10 p.

Anon., *An Approach to International Regulation of the Recovery of Deep Sea Ferro-Manganese Nodules,* The International Nickel Company of Canada, January 1973, 8. p.

[48]Pasho, D.W. and J.A. McIntosh, 'Recoverable Nickel and Copper from Manganese Nodules in the Northeast Equatorial Pacific—Preliminary Results', *Canadian Mining and Metallurgical Bulletin,* September 1976, p. 15-16.

[49]It is important to note in this regard that the possibility that the most readily exploitable portion of the resource can be controlled by a small number of companies or States could be increased considerably if the conditions associated with exploration permits are not carefully designed.

[50]Value resulting from "global estimator" approach — corresponding to the estimated area in which nodule grade exceeds 2.3% combined nickel and copper and abundance averages more than 6,500 dmt/km^2; Archer's "prime area" approach yields a value of 2.2 million km^2.

[51]U.S. Working Paper, 3 August 1970.

[52]If, following Archer's "prime area" approach, the potentially mineable seafloor area is only 2.2 million km^2, there would exist approximately 25 viable sites of 30,000 km^2 each and nearly 40 viable sites of 55,000 km^2.

[53]United States Senate Bill S. 1134, 93rd Congress.

[54]U.S. Draft United Nations Convention on the International Seabed Area, Working Paper 3 August 1970.

[55]See Reference 47.

[56]United Nations Third Conference on the Law of the Sea. A/CONF.62/WP.8/Part. I.

PART IV
COMMERCIAL CONSIDERATIONS

CHAPTER IX

1. PROBLEM ADDRESSED AND A SUMMARY OF THE DISCUSSION

1. Problem Addressed

The extent to which any of the deposits of manganese nodules could be considered as part of the world resource base of the metals they contain is entirely a question of present and future mineral economics. Tinsley's paper entitled "Manganese Nodule Mining Industry: A Study of Expected Investment Requirements", was therefore an important contribution to the Group of Experts meeting. In introducing the paper, Tinsley touched briefly on the impacts that the production of nickel, copper, cobalt and manganese would have on world markets and concluded that for nickel and copper they would be small, although the impact of cobalt could be more significant. The discounted cash flow method of financial analysis, although most appropriate, was not used as it was sensitive to the timing of capital investment, a factor that was difficult to predict from outside the companies involved. Tinsley emphasized the uncertainties associated with values adopted for calculation purposes and stressed that these values could have an inaccuracy of at least ±30%. Each of the factors that influenced both cost and revenue were described. He concluded that for the one million metric-ton-per-year model mine, which he had examined, the net return on investment after tax might vary between 4.4 and 6%.

Welling discussed the composition of the consortia and outlined the work that was being undertaken and the steps that would have to be taken before a full scale mining system could become operational, allowing full-scale production to begin. He emphasized that, at present, all the groups were involved in research and development programmes rather than in mining development programmes. Although the technical problems that had to be overcome were formidable, he was confident that with a systematic approach and sufficient allocation of research funds, solutions would be found. It was clear, even at this stage, that whatever advances were made in the development of the technology involved, the recovery of metals from nodules would not be a low cost operation. After tests of mining systems, a prototype mining stage would be necessary before investment was made in a full scale system. Given the appropriate legal environment, full scale mining could be possible by the mid 1980's.

2. Summary of the Discussion

Tinsley, in presenting his paper, emphasized that his calculations should be interpreted as being only a representative 'base case' for the purpose of stimulating discussion. He attempted to make his analysis as explicit and detailed as possible, in order to allow the experts to focus on specific items of cost and revenues; he felt that this was more valuable than a highly aggregated analysis which kept the details and assumptions hidden in the background.

It was felt by some experts that while Tinsley's figures required significant modification to reflect the current situation, they were a closer approximation of the current situation than most other estimates. If, at the conclusion of research and development programmes and at the moment of the investment decision, the economies of the mining venture proved to be similar to Tinsley's figures, then no consortium would commit several hundred million dollars to the mining programme.

Specifically, it was noted that a three-product operation would not be at all economical at the scale discussed by Tinsley, i.e. one million tons-per-year. The cost specified for the mining vessel was felt to be too low; the assumed speed of the bottom miner was too high and the width of the head too large. These factors would affect the assumed recovery rates and hence would change both costs and revenues. Tinsley included no estimate of working capital requirements for his representative project, which was an important omission. In addition, there was no overdesign in the capacity of the mining system, which would be necessary to make up for the instances when the recovery rates fell below rated levels due to weather and mechanical problems. The inclusion of an overdesign element would increase capital costs, and it also would affect operating costs, about which there was general uncertainty in any case.

The low cash flow of the representative project was noted, and it was pointed out that any system of investment credits or depletion allowances would obviously increase the cash flow. The rate of return projected by Tinsley was felt to be far too low for such a venture, and it was felt that the initial ventures might require at least a rate of return on investment in the range of 25-30 per cent. The mining company would want to know, as accurately as possible, the cost structure of nodule production, its evolution over time, and its place in the company's hierarchy of the production costs for its various mines.

There was no doubt that, in the long-run, metal production from nodules would be an economic means of providing raw materials. Furthermore, developments in technology would almost certainly improve the cost picture over time. The examples of taconite and low-grade porphyry copper mining in the United States were cited as analogies.

It was felt that ocean mining ventures might face problems in attracting capital because of the long gestation period during which bankers would expect sufficient coverage of debt. It was not clear upon which capital markets these ventures would draw, and there was no indication of likely mixes of debt and equity financing. The actual investment decision for first generation mining ventures would not be made before the mid-1980's, and therefore there would probably not be any significant production volumes before the late 1980's.

It was emphasized that at present, the industry was engaged in research and development and *not* in a programme of mine-site development for ocean mining. The current state of activities was geared to predicting engineering performance or operability. Examples given included determining the reliability of the mining head and pipeline under operational conditions. Additionally, it was emphasized that a prototype test was indeed a full scale test requiring a minimum threshold investment.

Below is a listing of the various consortia and groups involved in deep sea mining research and development:*

1.	**Ocean Mining Associates (O.M.A.) (a Virginia partnership)**	
—	Deepsea Ventures Inc. (a subsidiary of Tenneco Inc. (USA) and service contractor to the Consortium	
—	Essex Minerals Co (a U.S. corporation owned by United States Steel Corp) (USA)	33⅓
—	Sun Ocean Ventures (a U.S. corporation owned by Sun Co. Inc.) (USA)	33⅓
—	Union Seas Incorporated (a U.S. subsidiary of Union Miniere S.A. Belgium)	33⅓
2.	**Ocean Management Incorporated (O.M.I.)**	
—	International Nickel Company of Canada Ltd. (INCO)	25
—	SEDCO Incorporated (USA)	25
—	A.M.R. Group of the Federal Republic of Germany[1]	25
—	The Deep Ocean Mining Company of Japan (DOMCO Ltd.)[2]	25
3.	**Ocean Minerals Company (O.M.C.)[3]**	
—	Ocean Systems of Lockheed Missiles and Space Company Inc. (USA)	
—	Amoco Minerals Co. (a subsidiary of Standard Oil Co. of Indiana) (USA)	
—	Billiton International Metals B.V. (a subsidiary of the Royal Dutch/Shell Group (Netherlands))	
—	Bos Kalis Westminister (Netherlands)	
4.	**The Kennecott Group**	
—	Kennecott Copper Corp. (USA)	50
—	Rio Tinto Zinc Deep Sea Enterprises Ltd. (U.K.)	10
—	British Petroleum Minerals Ltd. (U.K.)	10
—	Consolidated Goldfields (U.K.)	10
—	Noranda Mines Ltd. (Canada)	10
—	Mitsubishi Corp. (Japan)	10
5.	**Association Francaise pour l'Etude et la Recherche des Nodules (AFERNOD (France))[3]**	
—	Centre National pour l'Exploitation des Océans (CNEXO)	
—	Commissariat à l'Energie Atomique (CEA)	
—	Bureau de Recherches Géologiques et Minières (BRGM)	
—	Société le Nickel (SNL)	
—	Chantiers de France-Dunkerque (from the Empain Schneider Group)	
6.	**The Continuous Line Bucket Group (CLB)[3]**	
—	Phelps-Dodge Corp. (USA)	
—	Dome Exploration (Canada)	
—	Occidental Minerals Corp. (USA)	
—	Superior Oil Co. (USA)	

— Broken Hill Proprietary Co. Ltd. (BHP), (Australia)
— CNEXO (France)
— Utah International Inc. (USA)
— DOMCO Group (Japan)
— Placer Development Ltd. (Canada)
— Teek Corp. Ltd. (Canada)
— COMINCO Ltd. (Canada)
— INCO, Ltd. (Canada)
— AMR Group (Federal Republic of Germany)
— Sumitomo Heavy Industries
— Atlantic Richfield Co. (USA)
— United States Steel (USA)
— Sociētē Le Nickel (France)
— General Crude Oil (USA)
— Noranda Mines Ltd. (Canada)

---

7. **Deep Ocean Mining Association (DOMA)**[3] (Japanese)
— C. Itoh and Co. Ltd.
— Furukawa Mining Co. Ltd.
— Hitachi Shipbuilding and Engineering Co. Ltd.
— Ishikawajima-Harima Heavy Industries Co., Ltd.
— Japan Metals and Chemicals
— Kanematsu-Gosho Ltd.
— Kawasaki Heavy Industries Ltd.
— Kawasaki Steel Corp.
— Kobe Steel Ltd.
— Marubeni Corp.
— Mitsubishi Corp.
— Mitsubishi Heavy Industries Ltd.
— Mitsubishi Metal Corp.
— Mitsui and Co. Ltd.
— Mitsui Mining and Smelting Co. Ltd.
— Mitsui Shipbuilding and Engineering Co. Ltd.
— N. Y. K. Line
— Nichimen Company Ltd.
— Nippon Electric Co. Ltd.
— Nippon Kokan Kabushiki Kaisha
— Nippon Mining Co.
— Nippon Steel Corp.
— Nippon Yakin Kogyo Co., Ltd.
— Nissho-Iwai Co., Ltd.
— Nittetsu Mining Co.
— Pacific Metals Co., Ltd.
— Sumitomo Metal Industries
— Sumitomo Metal Mining Co., Ltd.
Sumitomo Shipbuilding and Machinery Co., Ltd.
— Sumitomo Shoji Kaisha Ltd.

— The Dowa Mining Co., Ltd.
— Victor Company of Japan, Ltd.

8. **Eurocean**[3]
— Peachiney Ugine Kuhlmann (France)
— Boliden (Sweden)
— Granger International (Sweden)
— Kockums McKaniska (Sweden)

*These tables were constructed on the basis of the most recent published literature on deep-sea nodule consortia and groups. In particular, the publications of the United States Senate Committee on Interior and Insular Affairs were utilized extensively.

[1] The A.M.R. group consists of Metallgesellschaft A.G., Preussag A.G., Rheinische Braunkohlenwerke A.G. and Salzgitter A.G.

[2] DOMCO Ltd. represents six firms from the SUMITOMO combine, Ataka and Co Ltd., Toyo Meuka Kaisha Ltd., Marubeni Corp. Kyokuyo Co. Ltd., Dowa Mining Co. Ltd., Nijuson Mining Co. Ltd., Shinko Electric Co., Nissho-Iwai Ltd., Tokyo Rope Manufacturing Co. Ltd., and Mitsui OSK Alliance Ltd.

[3] The exact percentage ownership is unknown.

CHAPTER X

MANGANESE NODULE MINING INDUSTRY: A STUDY OF EXPECTED INVESTMENT REQUIREMENTS

1. Introduction

Much of the data on the occurrences of manganese nodules remain in the category of "indicated paramarginal resources" or "inferred, hypothetical, and speculative paramarginal resources."[1] The number of samples in the public domain is still small relative to the vast areal extent of the world's oceans and the sample density within the most closely studied North Equatorial Pacific zone is still widely spaced. However, good data presumably reside in the private sector on individual target mine-site areas. Nevertheless, industry authorities conclude that the nodule abundance and grade is quite variable within these target zones.[2] Preliminary analysis has shown that 20-25% of a target nodule mine-site may be unmineable because of obstructions and that high nodule populations are likely to occur in the vicinity of these obstructions.[3]

The total resource base is not well understood and has been widely misstated. A commonly quoted figure is that of Mero[4] and is usually rounded off to 1.5 trillion tons of nodules. Actually, Mero computed a total of 1,656 billion tons of nodules for the Pacific Ocean based on average concentration over the whole area, but hastily added: "Most of the calculations concerning manganese nodule tonnages are of the nature of speculation." More recent studies by arithmetic, quasi-statistical, and statistical methods point to the conclusion that the number of mine-sites in the Pacific Belt and elsewhere may be somewhere between twenty[5] and two hundred.[6] A mine-site is usually regarded as having 60-80 million tons of mineable manganese nodules from a surface abundance of 5 kg/m^2 to 10 kg/m^2. This represents a tonnage figure for the *total world's* nodule mine-sites of approximately one two-hundredth (0.5%) of Mero's *Pacific Ocean* calculation. These "orders of magnitude" differences in the various nodule tonnage estimates signal that much caution is required in extrapolating from what can only be regarded as preliminary estimates in an analytical regime which depends on relatively sparse data of questionable statistical validity. Although some will argue that the nodule resource is vast—and this is a popular conception—one must be aware of the data limitations.

The highly variable abundance serves to highlight just one of the technical risk components which will mitigate against manganese nodule mining becoming a bonanza business.[7] It will be, in fact, a very risky undertaking and prospective mining companies have alternative investment opportunities to pursue.[8] Furthermore, the enormous investment required for each venture will, among other factors, act as a brake on actual investment, particularly for the first nodule mines.

The purpose of this paper is to present a theoretical analysis of nodule mining investment requirements, given explicit assumptions which can hopefully serve as a framework for discussion. The results, which are presented as hard numbers, are no more than informed estimates adapted from published sources and represent one case in a range of possible analyses. However, sufficient detail is presented to generate sensitivity analyses. This paper must not be characterized as a determination of the economic feasibility of manganese nodule mining.

In this kind of analytical approach one must hope for compensating errors, and even discussions with other experts cannot narrow down the margin for error in individual parts of this analysis due to the proprietary nature of most of the economic studies. Further, it must be remembered that the parameters under scrutiny are for an as yet unformed industry for which there is no historical data.

This paper follows a three part structure. First, the technical requirements for nodule recovery and processing are summarized and, secondly, in the light of these developments, a postulated nodule mining industry is analyzed in terms of its investment requirements. Thirdly, the capital and operating costs are modelled for a small, 1 million ton per year nodule mine. Supporting figures and tables are attached.

2. Nodule Recovery and Processing

2.1 Mining

The mining systems under development have four major components:

(i) mining head/nodule collector
(ii) materials elevation/hoisting
(iii) surface vessel/mining ship
(iv) product transportation to shore plant

2.1(i) Nodule Collector

The heavy research and development (R&D) burden on the development of nodule mining and processing—expected to range between $50 million and $100 million per venture—virtually mandates that large tonnages be processed with the resultant economies of scale required to make a project viable. Early calculations indicated that an optimum mining rate would be approximately 1.3-1.5 million tons of nodules per year.[9] Work has since led to the consideration of large tonnage throughputs, such as 3 million tons per year or more although there is some debate on the question of whether one or two mining ships is preferable. The exact tonnage chosen by any venture will, of course, be optimized for the various inputs chosen.

The production levels will require a wide (up to 30m), unwieldy collector operating on the deep-seabottom which can vary in physical characteristics from a sticky grease to extremely firm surfaces, exhibiting brittle features. Numerous patents have been issued for nodule gathering devices which are analagous to conventional agricultural harvesting or fishing devices and to adaptations of wet-terrain vehicles.[10 11]

A considerable amount of collector testing has been conducted by various groups both in on-land testing tanks and at sea.[7 12] Current and future investigations tend toward a simple, towed collector, although more complicated "robot" devices are under study.[13]

2.1(ii) Materials Elevation

There are three contenders in the method to elevate gathered nodules some 4000-5000 m to the ocean surface: airlift, hydrolift (hydraulic hositing), and the continuous line bucket (CLB). The airlift has been tested twice at water depths of 900 m by Lamotte[12] and 5000 m[14] and a further test was scheduled for late 1977-early 1978 on a one-fifth scale. However, the 30-50% efficiency of an airlift is a disadvantage.

A promising technique appears to be the hydrolift which offers predictable, efficient results. A considerable amount of on-land test loop work has been conducted and a hydrolift test was scheduled for late 1977-early 1978 on a one-fifth scale and may be tested on a fullscale basis in 1979. The technological developments required are the placement of the various pumps along the pipe length, the engineering of 5000 m long, wide-diameter pipe strings, and pipe handling equipment. The main disadvantage of a hydrolift is the placement of moving mechanical parts at depth, whereas for the airlift the air compressors can be mounted on the deck of the surface vessel.

The CLB system uses buckets attached to a long, continuous, buoyant rope loop which is driven by ship-board traction motors. A single ship test has been conducted[15] and a two-ship test is in the planning stages[7], according to Tinsley.

2.1(iii) Mining Ship

The first mining ships will be similar to a self-propelled, dynamic positioned drilling ship but with (a) surge storage capacity for nodules of about a week, (b) capability to transfer the nodules to a bulk or slurry ore carrier, and (c) enlarged accommodation for for the crew. The design will incorporate the requisite power pipe handling equipment. The hull could be a modified ore carrier. The mining ship may be required to support some preliminary at-sea processing. [16] [17] [18]

Future mining platforms may include semi-submersibles which could be powered by Ocean Thermal Energy Conversion (OTEC) power plants.[19] The favourable OTEC generation belt coincides with much of the North Equatorial manganese nodule zone. It is further possible that such an OTEC plant could generate the ammonia required to process nodules for currently proposed hydrometallurgical flow-sheets.[20]

2.1(iv) Ocean Transportation

Conventional bulk, slurry, or ore-bulk-oil (OBO) vessels will be used for ocean transportation with provision for loading-at-sea. Vessels will probably be greater than 30,000 tons.[21] Slurry transportation would be compatible with port-to-process plant slurry pipelines.

2.2 Processing

Extensive process research has narrowed the number of process routes for the extraction of nickel, copper, cobalt, manganese, molybdenum, and zinc from the recovered nodules.[22] These can be summarized as follows:

(i) Pyrometallurgical: Smelting with flux and sulfur source to produce a high-manganese slag and a metal concentrate.[23]

(ii) Hydrometallurgical: Cuprion process using an ammonical leach in a reducing atmosphere and recovery of nickel and copper by solvent extraction-electrowinning (SX-EW) with separate cobalt-molybdenum recovery.[20]

(iii) Hydrometallurgical: Sulfur dioxide roast with water leach to yield cement copper with nickel and cobalt recovered by hydrogen reduction.[24 25]

(iv) Hydrometallurgical: High-pressure sulfuric acid leach with SX-EW.[22]

(v) Hydrometallurgical: Chloride leach with SX-EW and high-purity manganese recovery.[22]

Process testwork continues worldwide from bench-scale sizes to continuous, small pilot-plants. Once substantial tonnages have been recovered from the prototype mining tests, larger scale pilot plants in excess of 10 tons per day will be used to validate the process route selection. As a rule, the pyrometallurgical route will yield a manganese product suitable for conversion to ferromanganese. Manganese recovered from a hydrometallurgical plant will likely be of a higher manganese content than ferromanganese (78% Mn).

3. Industry Financial Requirements

General assumptions for this section:

North Pacific Nodules
Water Depth 12,000 feet

Grade: 1.3% Nickel
.25% Cobalt
29% Manganese

Venture 1 onstream in 1984.
Venture 2 onstream in 1985.
Venture 3 onstream in 1985.
Venture 4 onstream in 1986.

Total Investment in escalated dollars of $2.6 billion excludes R&D, Working Capital, and Exploration.

This part of the paper assumes an optimistic "best-case" scenario for the first nodule mines. The assumed investment schedule allows little or no slippage in the design and testing programmes now underway by five groups presently investigating manganese nodule mining. This optimistic scenario is somewhat tempered by the fact that most of the technology is known for each sector of nodule mining, transportation, and processing stages[10 22 24] However, the ability of any sector to function within each fullscale, totally integrated stage is being tested in prototype operations conducted in 1977-1978 for the at-sea mining[14 26] and in land-based pilot plants during 1977-1979 for the processing stages, according to reports in *Ocean Industry*.

A further key assumption is the institution of an equitable investment regime by 1978 which will provide a stable legal and economic climate for each nodule mining venture. It is not the purpose here to discuss the on-going Law of the Sea Conference but rather to present a possible scenario given the most favourable conditions.

Given these assumptions and the technical outline above, it can be postulated that a four-venture industry would proceed forthwith to the development of medium-to-large-scale nodule mines and associated processing plants. The schedule assumed must be regarded as "ambitious" and presumes a decision taken within the next few years for an all-out effort to get into production by each of the four ventures.

In this paper, it is postulated that one venture may come onstream in 1984, two in 1985, and one in 1986. It must be stressed that the postulated scenario reflects what could happen under the most optimistic of circumstances. It must not be regarded a forecast or projection.

Excluding R&D, working capital, and exploration costs, the total requirements in 1988 are $2.6 billion on an escalated cost basis (Figure 1, Tables 1 and 2). The R&D requirements for each ventures will be at least $50-$100 million. The exploration cost would total at least $2 million per year for each venture. The total capital requirements for the four ventures easily exceeds $3 billion by 1988.

An annual nodule production of 8,300,000 tons per year by 1988 would yield approximately 98,000 tons of nickel, 82,000 tons of copper, 14,000 tons of cobalt, and 950,000 tons of manganese which would represent 9%, 1%, 27%, and 7% of the projected 1990 world requirements for the four metals, respectively (Table 3). The four-venture industry would have a very small impact on the world's mineral markets in 1990, except for cobalt. In general, metal production from nodules gets lost in the noise of the various forecasts/projections and a one percentage point change in demand growth rates reduces the nodule mining impact by 1990.

TABLE 1.
Estimated Capital Investment* Schedule
(constant 1977 $ million)

	80	81	82	83	84	85	86	87	88	
Venture 1		100	150	150	100	10	10	10	10	540
Venture 2	40	50	50	40	9	9	9	9	9	225
Venture 3			60	160	220	20	10	10	10	490
Venture 4	30	40	50	50	50	40	8	8	8	284

*Excludes R&D, Working Capital, Exploration

TABLE 2.
Estimated Capital Investment* Schedule
(escalated $ million)

	80	81	82	83	84	85	86	87	88	
Venture 1		144	233	252	181	20	21	22	23	896
Venture 2	53	72	77	67	16	18	19	20	21	363
Venture 3			93	269	398	39	21	22	23	865
Venture 4	40	58	77	84	91	77	16	18	19	480
Total Investment*		274	480	672	686	154	77	82	86	2604

*Excludes R&D, Working Capital, Exploration.
Annual Escalation: 10% to 1980, 8% to 1985, 6% thereafter.

TABLE 3.
Annual Production Summary
(000 tons)

	Mine	Nickel	Copper	Cobalt	Manganese	
Venture 1	2,700	31.8	26.9	3.7	——	Hydrometallurgy
Venture 2	1,000	11.3	9.2	2.2	253	Hydrometallurgy
Venture 3	3,000	35.0	29.7	5.3	695*	Pyrometallurgy
Venture 4	1,600	19.5	16.1	2.4	——	Hydrometallurgy
	8,300	98.0	82.0	14.0	950	
Total, as Percentage of Requirements by 1990						
U.S.		35%	2%	110%	57%	
World		9	1	27	7	

*Manganese content in ferromanganese

FIGURE 1. Nodule Mining Investment

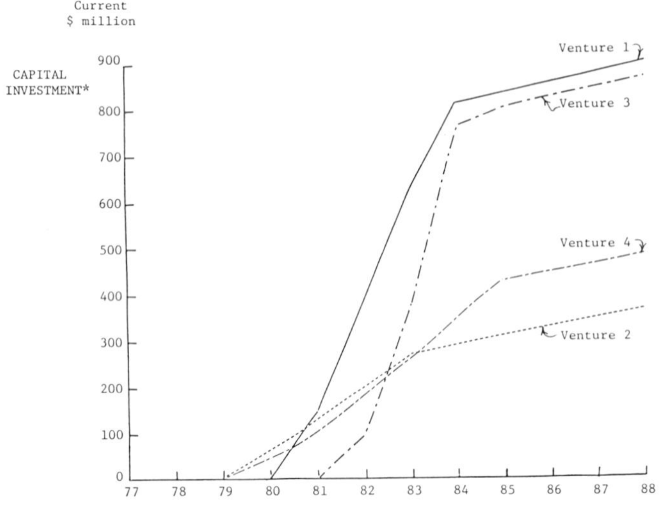

Cobalt is the smallest tonnage metal produced. It is a substitute for nickel in certain limited markets, particularly for electroplating. However, present United States demand for electroplating nickel alone is almost twice the total nodule cobalt tonnage assumed produced in 1988. Nodule cobalt would signficiantly penetrate the world cobalt market, but the tonnage of available cobalt would still only amount to about 5% of the projected world nickel market. It is likely that cobalt will be readily able to expand its markets by that time, and not just for electroplating, but for other functional applications as well.

The impact of nodule production on the individual metal markets are shown for the world and the United States in Figure 2-9. This topic itself would be the basis of another study.

FIGURE 2. Impact of Nodule Nickel Production on US Nickel Demand.

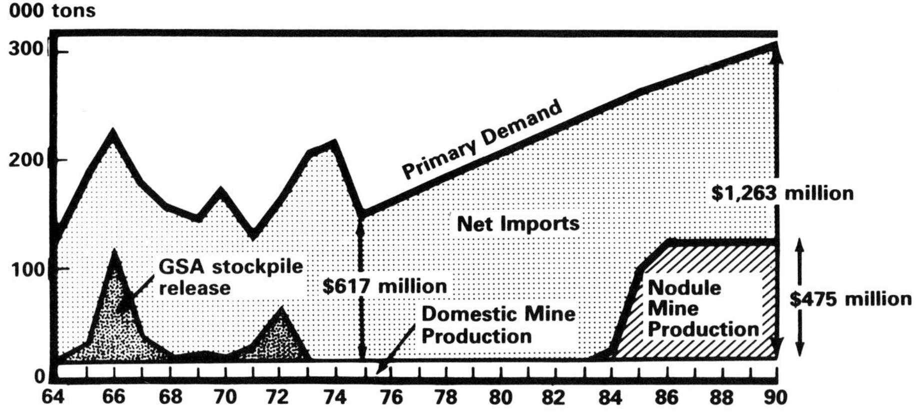

Source: USBM, Minerals and Materials, August 1977
Forecasts-USBM, Mineral Facts and Problems, 1975
Dollars are constant $1975. Nickel Price is $2.03 per lb.

FIGURE 3. Impact of Nodule Nickel Production on World Nickel Production.

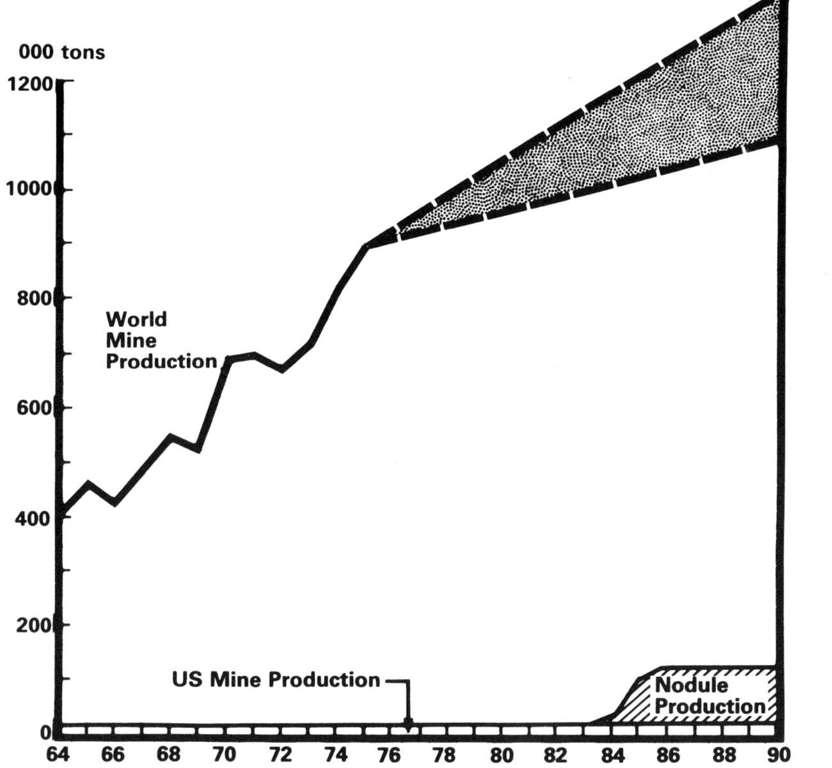

Source: USBM, Minerals in the US Economy 1964-1975
Forecasts-Various

FIGURE 4. Impact of Total Nodule Copper Production on US Copper Consumption

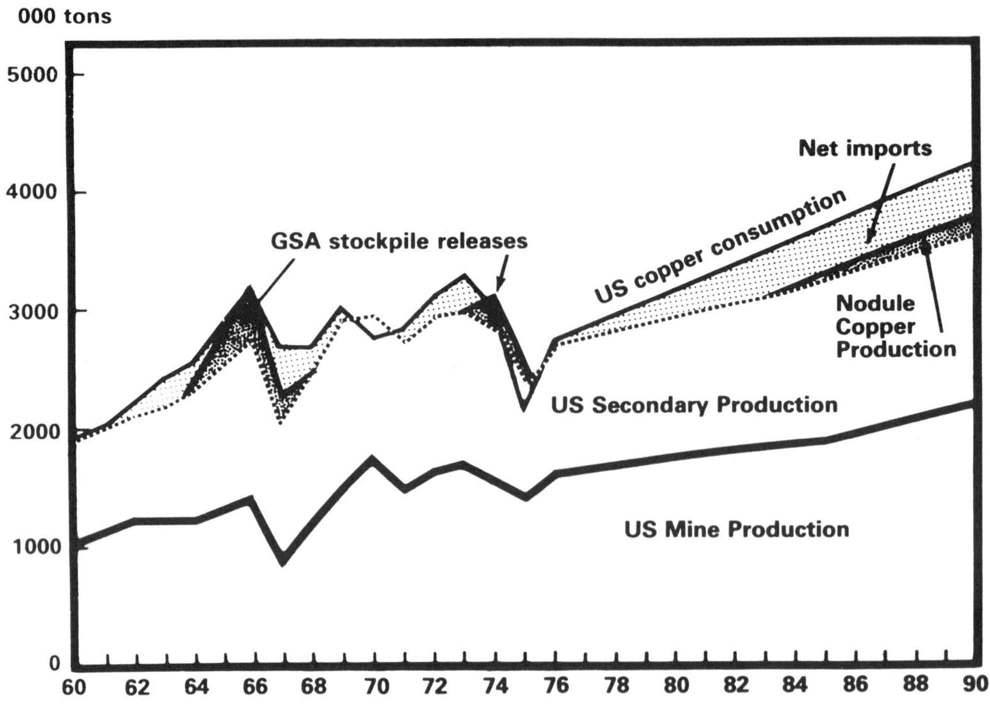

Source: USBM, Minerals and Materials, August 1977
Forecasts-Various

FIGURE 5. Total Nodule Copper Production Relative to World Copper Mine Production.

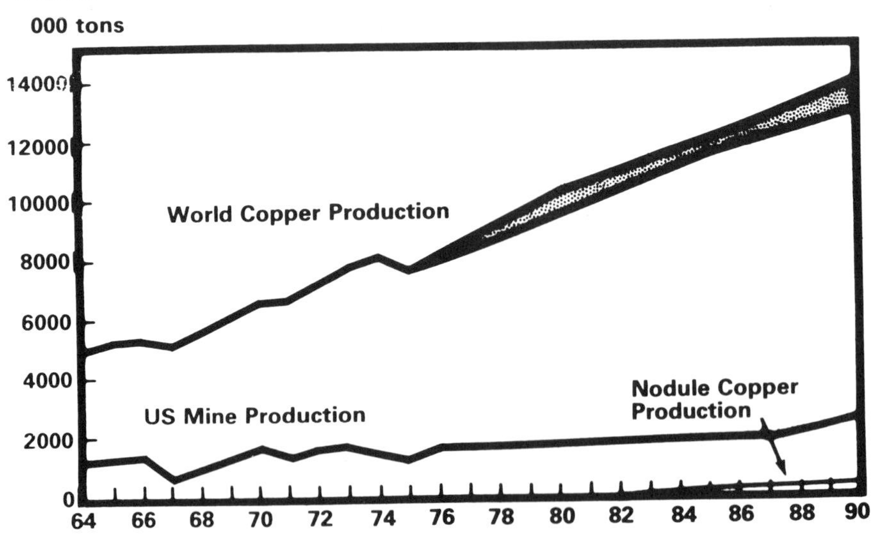

Source: USBM, Materials in the US Economy, 1964-1975
Forecasts-Various

FIGURE 6. Total Nodule Cobalt Production Relative to US Consumption.

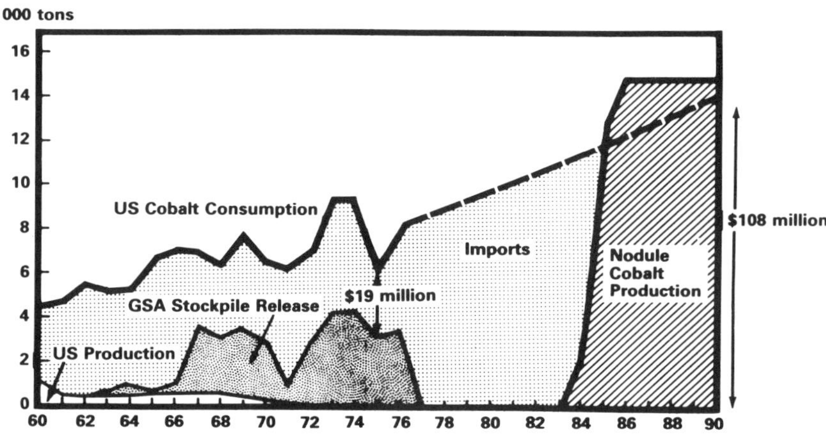

FIGURE 7. Impact of Total Nodule Cobalt Production on World Cobalt Production

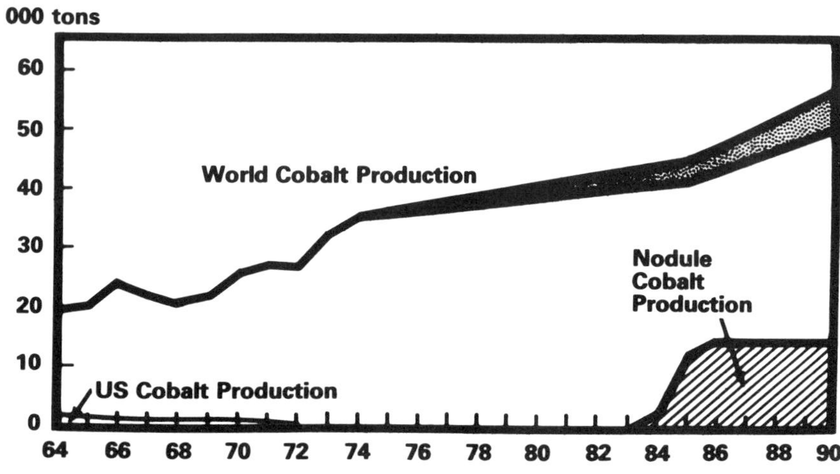

FIGURE 8. Impact of Nodule Mining on US Manganese Imports.

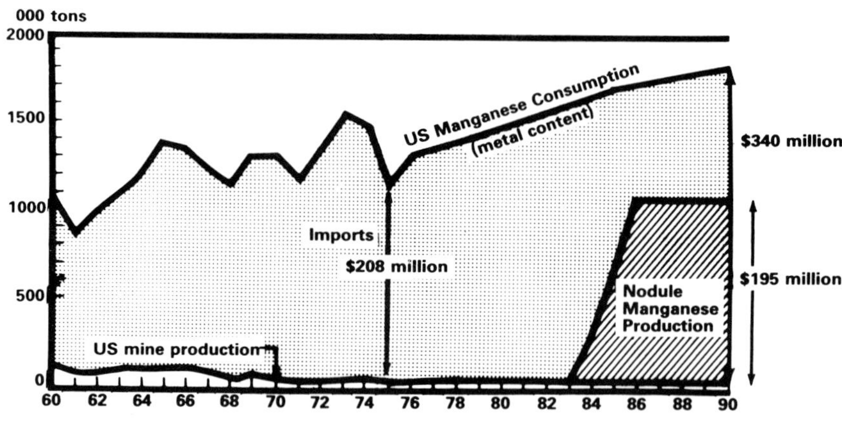

FIGURE 9. Impact of Nodule Mining on World Manganese Production.

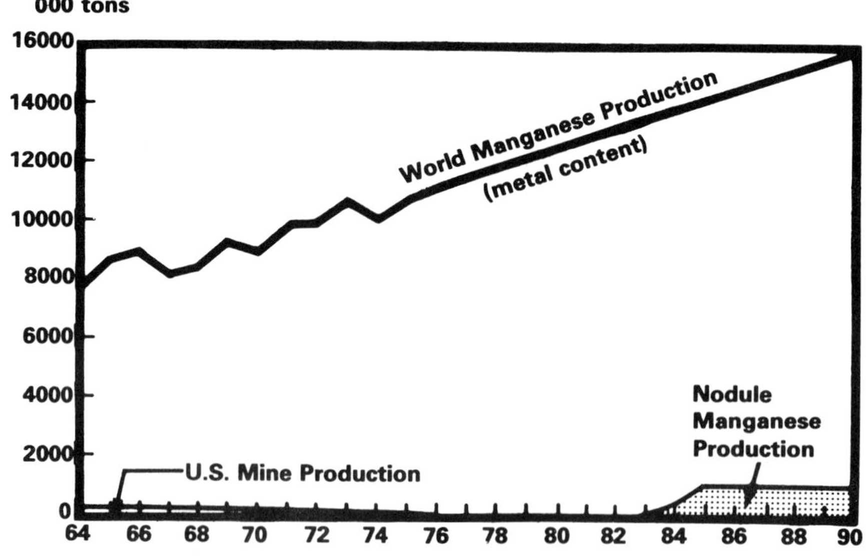

4. Financial Analysis — 1 Million Ton Mine

General assumptions for this section:

Mine: 1.3 million wet metric tons per year
1.0 million dry mtpy (metric tons per year)
4,700 wet mtpd (metric tons per day); 3,600 dry mtpd

Operations: 300 days per year
Maneuvering loss: additional 20 days per year
Net operations: 280 days
Abundance: 10 dry kg/m^2

		Nickel	Copper	Cobalt
Net Mined Grade:		1.3 %	1.0	0.2
Prices:	1976:	2.28	0.75	4.30
(1976 $ per lb.)	1985:	2.49	0.82	4.60
	2000:	2.49	0.88	2.64

Much analysis exists in the literature on nodule economics[27 28 29] but the overall capital and operating costs are grossly aggregate numbers. Furthermore, compilations of these estimates suffer from changes in costs over time (e.g. 1968 vs. 1973 estimates)[30] as well as an absence of explicit assumptions in many instances.

The following is a detailed outline of a hypothetical 1-million dry ton per year mine from which nickel, copper, and cobalt are recovered, but not manganese. A converva-tive estimate was made of each input variable based on published or interpolated data. For capital costs, a six-tenths exponent rule was used uniformly to modify capacity estimates. All costs were brought to a common early 1976 dollar base using Marshall & Stevens indices, wholesale price indices, or other published estimates.

4.1 Exploration and R&D

The mine grade is assumed constant over the mine life. Exploration costs of about $2 million per year are not included and R&D costs are capitalized. R&D funds are roughly evenly divided between (i) pilot and test mining and (ii) bench- and pilot-scale process testing, culminating in a final 25-50 mtpd.

4.2 Process Plant

Two processes were examined in parallel to enable comparisons between different capital and operating costs assumptions. A compariatively higher capital cost adaptation of the Nicaro ammonia leach yielded lower plant operating costs, whereas a lower capital intensive sulfur-oxide roast yielded higher plant operating costs.[22 24] In general, the processes can be described as follows:

4.2.1 Ammonia leach — Crush

Partial dry in fluid bed or rotary dryer; grind to minus 20 mesh; roast at 1100° F in reducing gas; leach with ammonia-ammonium carbonate; recovery soluble amine metal complexes by air sparging followed by counter-current decantation; solvent extraction (SX) of copper, nickel, and cobalt; electrowin copper and nickel; precipitate cobalt. Ammonia recovery after washing.

4.2.2 Sulfur oxide roast — Crush:

Partial drying; grind to minus 20 mesh; roast at 600-700° in air-sulfur dioxide atmosphere; water leach; thicken to reject tails and iron; recover cement copper on iron scrap; precipitate nickel, cobalt, and remaining copper with hydrogen sulfide at 120°C under 50 psig; dissolve precipitate in sulfuric acid; clairfy; solvent extraction-electrowinning (EW) of nickel and cobalt. No zinc sulfate credit assumed. Sulfur dioxide gas concentration roughly equivalent to off-gas from pyrite roasting. Credit from exothermic leach process.

4.3 Mining Ship — Ocean Transportation

The choice of new or used vessels, vessel size, and flag of operation can drastically alter the cost estimates. Bulk carrier conversions are possible for mining platforms[32] but some of the unique duties of the mining ship suggest that ship construction may be necessary (Table 4).

TABLE 4
Capital Cost Estimate for Mining Ship
(1976$)

Build modified bulk carrier	
50,000 dwt (platform)	$19 million (see Figure 12.)
Modification	8
Airlift Equipment	3
Pipe	3
Navigation, etc.	5
Living quarters	4
50,000 dwt mining ship	$42 million

It is assumed that suitable ore-bulk-oil or bulk carriers would be chartered for the 1600 mile trip from the mine site to the port. Figure 10 and Table 5 show the cost comparison of owned vs. chartered vessels. The average of the two is used in the study. A slurry transfer at sea is assumed. Note, however, that ocean freight rates are subject to upward pressures at times (Figure 11).

TABLE 5.
Estimate of Ocean Freight Cost For Manganese Nodule Mine (Vessel Ownership)
($000 per year)

	Wages	Ins.	Unin. Loss	Rep.	Prov.	Stor.	O/H	Cap. Rec.	Total	$/opr. day
Non-U.S.	400	360	50	250	35	80	90	1,300	2,665	8,540
U.S.	1,000	450	70	300	35	80	90	1,300	3,325	10,650

Assumptions:

50,000 dwt bulk carrier; 15 knots; 312 operating days
foreign flag

		Bunkers
Mileage one way	1,600	
Days steaming	9	40 t/d = 360 T
Days loading	10	15 t/d = 150 T
Days in port	8	10 t/d = 80 T
Voyage time	30	10 t/d = 590 T

Ship: 30 days	230,600	
Bunker: 590 T @ $80	47,200	
Port Charge:	30,000	
Voyage Cost	$307,800	= $6.84* per wet mt of nodules (1976$)

Cargo Capacity: 45,000 mt

*compared with $5.70 estimate under charter (see Fig. 11)

Other Estimates ($ per ton of nodules):

Sorenson & Mead	(1968)	5.80
Dorstewitz	(1971)	9.20
Mero	(1972)	4.00
Clauss	(1972)	6.20
Dubs	(1974)	4.00-7.00
Unpublished	(1975)	5.70

FIGURE 10.

OCEAN BULK TRANSPORTATION COSTS
4000 MILES (APPROX.) ONE WAY
COST PER METRIC TON (DELIVERED) 1976$
10
9
8
7
6
5
4
3
2
1
base
case
lower limit
50
0
20
40
60
80
100
120
140
160
SHIP SIZE (000) dwt

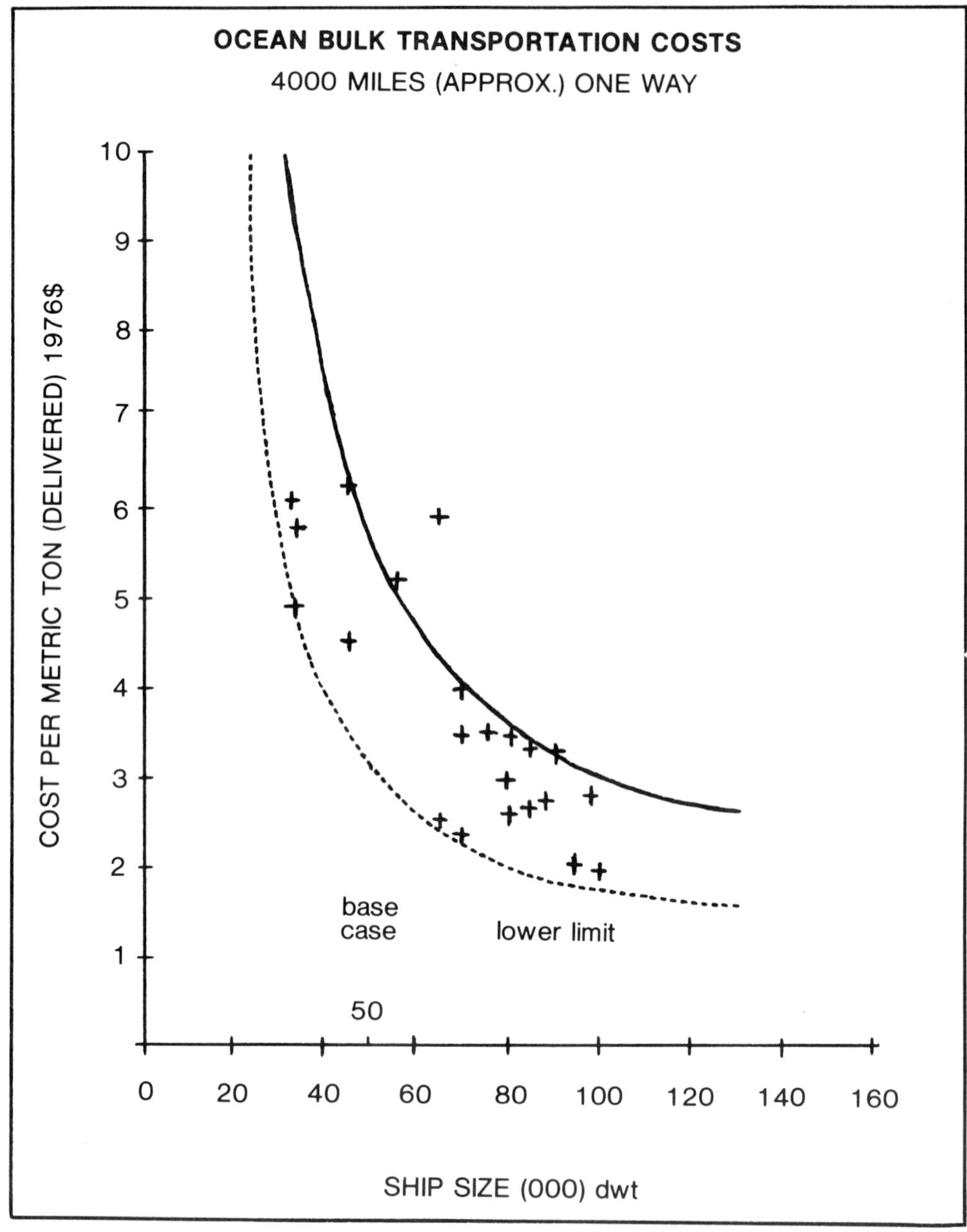

FIGURE 11.

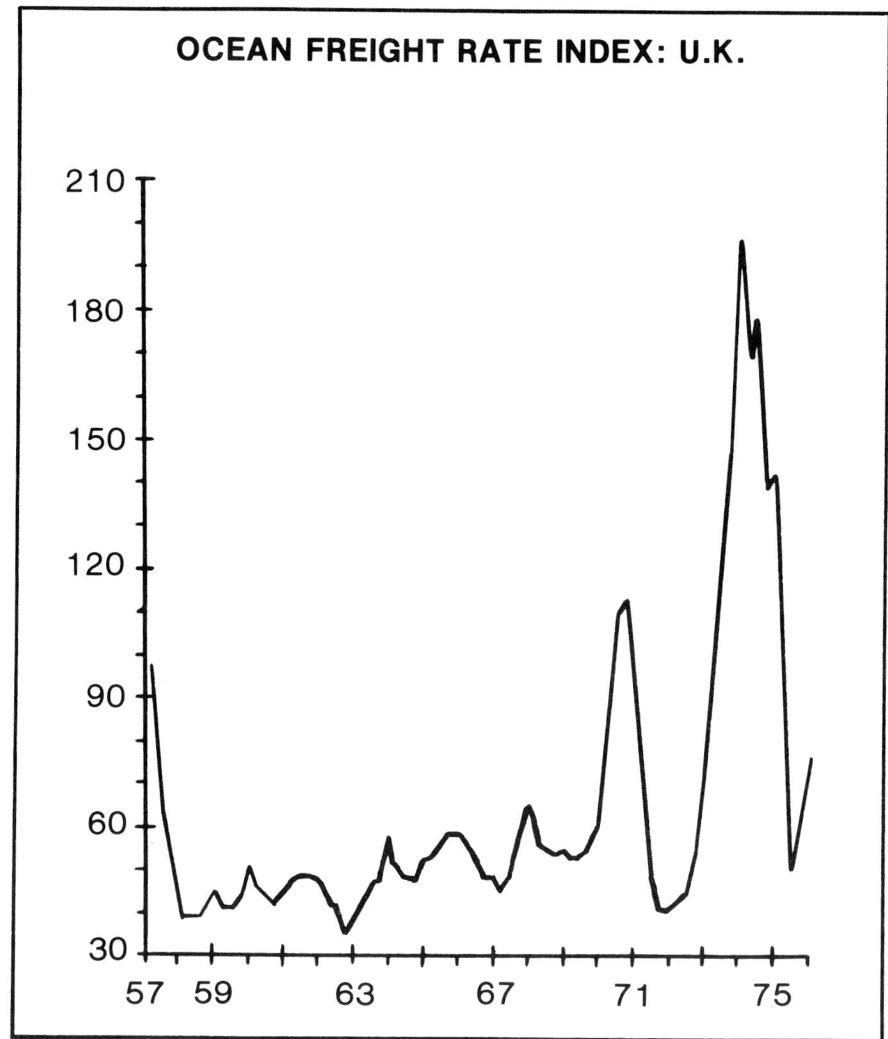

Source: International Monetary Fund

The mining ship is designed to have more than adequate storage for seven days production. A new bulk-carrier-equivalent cost is assumed for the hull, etc. (Figure 12) with additional investment for modification, mining equipment, and piping for hydrolifting (Table 4). A comparison of published hoisting energy requirements (Table 6) was used to determine the horsepower requirements for hoisting operating costs. Other horsepower estimates draw on Claus[9], for guidance on physical requirements.

The capital investment required for this small nodule mining venture ranges between \$230 million and \$270 million (Table 7). Total costs amount to \$69.05 and \$71.55 per ton for the ammoniacal and sulfur oxide processes, respectively (Table 8).

The background assumptions and calculations for Table 8 are shown in Tables 9 and 11 and to summarize (1976 \$ per day metric ton nodule feed):

FIGURE 12.

CAPITAL COSTS FOR NEW AND USED SHIPS
MILLION DOLLARS
20
10
8
6
5
4
Foreign construction
New bulk carriers
Delivery 1976-77
Used bulk
carriers
1975
OBO used
Foreign construction
Oil tankers
Delivery 1975
Used core/oil
carriers 1975
0 20 40 60 80 100 120 140 160 180 200 220
SHIP SIZE (000 dwt)

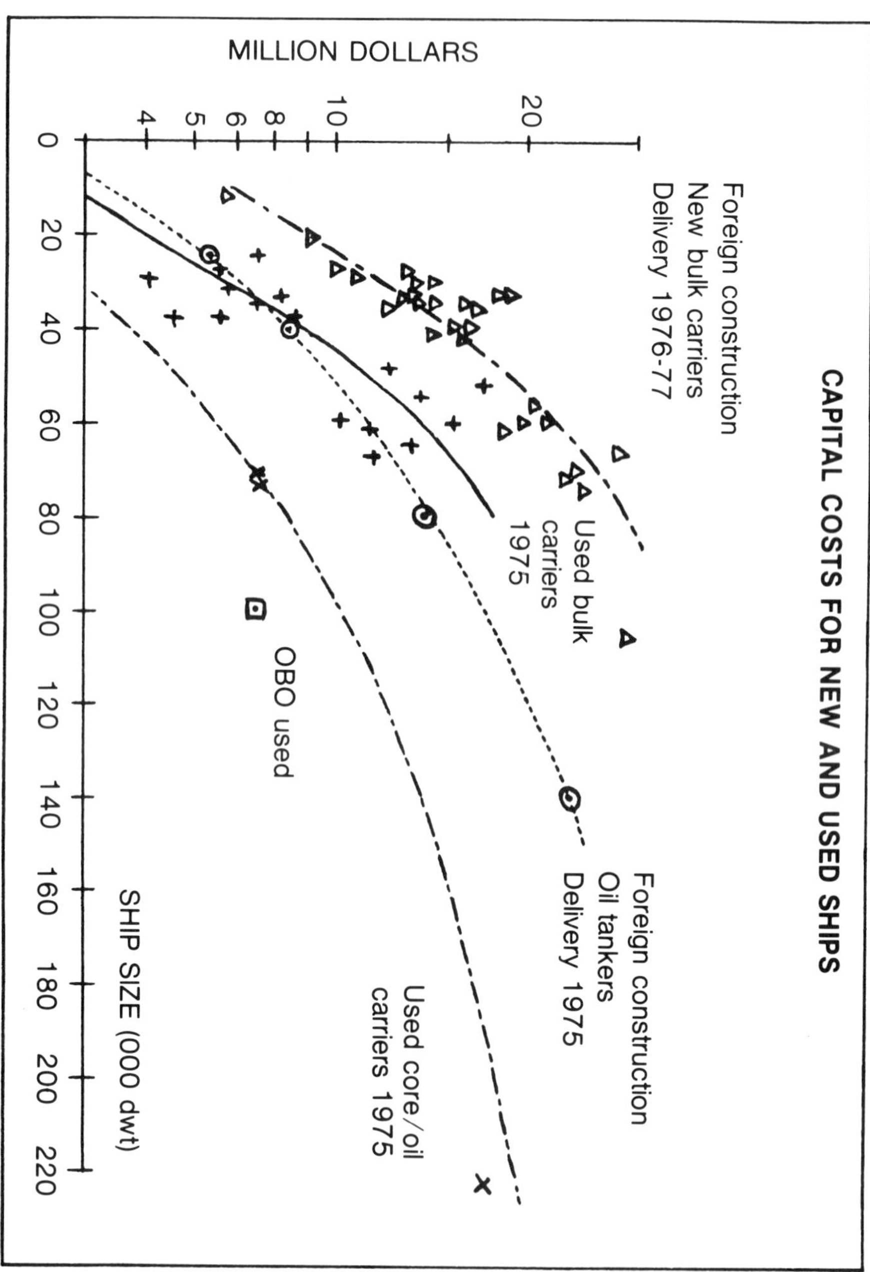

TABLE 6.
Comparison of Horsepower Calculations for Hydraulic Hoisting of Manganese Nodules

	Unit	Two-Phase Hydrolift		Three-Phase Air Lift			Three-Phase Light Media Lift
		Centrifugal Pumps					
Water Depth	m	14,500	9,000	5,500	11,000	14,500	9,000
Pipe Diam.	cm	51	51	38	38	51	30
Pipe Diam.	cm	51	51	38	38	51	30
Flow Vel.	m/sec	6.1	4.6	5.2	5.2	6.1	4.6
Solids	%	——	10	——	——	——	10
Production	tpd	5,511	9,000	6,500	6,500	5,511	5,800
Horsepower	h.p.	5,000	7,780	1,800	3,400	9,600	7,000
h.p.: tpd	ratio	0.91	0.86	0.27	0.52	1.74	1.21
Air Injection	m					3,000	
Reference:		16	4	38	38	16	39

TABLE 7.
1 Million mtpy Manganese Nodule Mine Summary of Capital Investment
(1976$ per dry mt of feed)

	NH*	SO*	Depr./ Amort. Yrs.
Mining Ship	$ 51	$ 51	15
Airlift and Pipe	6	6	3
Bottom miner	2	2	2
Port Facilities	4	4	15
Process Plant	115	80	15
R & D	62	56	15
Sub Total	$240	$199	
Working Capital	30	34	
	$270	$233	

*NH — Ammonia leach, SX-EW
**SO — Sulfur dioxide roast, Cement Cu, SX-EW

TABLE 8.
Total Costs Summary
(1976$ per dry mt of feet)

Process	NH	SO
Mining	$11.00	$11.00
Ocean Transportation	8.20	8.20
Dock & Transfer	1.10	1.10
Processing	21.35	27.20
Operating Costs	41.65	47.50
G&A	4.60	4.65
Depreciation	18.60	15.70
Amortization	4.20	3.70
Total Costs	$69.05	$71.55

Based on annual sales revenues in 1985 and 2000 which range from $86.95 to $91.77 per ton of feed (Table 12), a financial summary using averaging assumptions (Table 13) would indicate that the annual average cash flow as a percentage of investment equals 8.5-9.5% and the annual average net income after tax as a percentage of investment equals 3-4.5%. It must be stressed that these are average numbers and do not account for the time value of the investment or revenue stream under a conventional discounted cash flow analysis.

4.5 Conclusions

It can be concluded from this analysis that the level of return from the hypothetical small, 1-million-dry-mtpy nodule mine is less than many land-based mining ventures and that such nodule ventures would not yield bonanza after-tax income. Economies of scale will improve the return and it would appear that a 1-million ton per year nodule mine would be economically submarginal given the assumptions used in this analysis.

Obviously, since various groups continue to commit funds to nodule mining and processing R&D, it can be assumed that efforts to lower costs will be successful. In particular, the shift in emphasis toward larger nodule ventures is consistent with the findings in the last section of this paper. Furthermore, R&D work will likely provide significant improvements in various applications of the known technologies of nodule mining and processing. It can also be assumed that somewhat higher grade zones may be mined in the earlier years in order to improve the rates of return.

The analysis of an individual economic model for a small nodule mining operation does not provide information on the comparative economics with land-based production of these metals which would be included in a fully integrated economic model. This has important implications when one judges the rate of entry into the nodule mining industry (as hypothesized in section 3 above) and its cost competitiveness with other sources of nickel and cobalt, in particular. It was assumed for that section that the four ventures would have rates of return which would be more than double the rates derived from the mine model in section 4 with further leverage possible through financing a portion of the total investment requirements.

Notes and References

[1]Holser, A.F., *Manganese Nodule Resources and Mine Site Availability*, Ocean Mining Administration, U.S. Department of the Interior, May 1976, 12 p.

[2]Flipse, J.E., M.A. Dubs and R.J. Greenwald, 'Pre-production Manganese Nodules Activities and Requirements', *Mineral Resources of the Deep Seabed,* Hearings before the Subcommittee on Minerals, Materials, and Fuels of the U.S. Senate Committee on Interior and Insular Affairs, 15 March 1973, p. 602-700.

[3]Siapno, W.D., 'Exploration Technology and Ocean Mining Parameters', *American Mining Congress Convention,* San Francisco, 28 September-1 October 1975, reprint 24 p.

[4]Mero, J.L., *The Mineral Resources of the Sea,* Elsevier Publishing Company, Amsterdam, 1965, p. 174-175.

[5]Bastien-Thiry, H.B., J.P. Lenoble and P. Rogel, 'French Exploration Seeks to Define Mineable Nodule Tonnages on Pacific Floor', *Engineering and Mining Journal,* July 1977, p. 86-87, 171.

[6]Dubs, M.A., Testimony before Subcommittee on Minerals, Materials, and Fuels of the U.S. Senate Committee on Interior and Insular Affairs, 7 November 1975, p. 14.

[7]Tinsley, C.R., 'Nodule Miners Ready for Prototype Testing', *Engineering and Mining Journal,* January 1977, p. 81.

[8]Stern, M., Testimony before Subcommittee on Public Lands and Resources of the U.S. Senate Committee on Energy and Natural Resources and the Committee on Commerce, Science and Transportation, 20 September 1977, p. 2.

[9]Clauss, G., 'Theoretical and Experimental Investigations of Deep Ocean Mining Systems and their Economic Implications', *Second International Ocean Development Conference,* Tokyo, Japan, 5-7 October 1972, p. 1940.

[10]Li, T.M., and C.R. Tinsley, 'Deep Ocean Floor Nodule Mining — First Generation Techniques are here', *Mining Engineering,* April 1975, p. 47-52.

[11]Pasho, D.W., 'A Qualitative Consideration of some Mining Machine Seafloor Interactions', *Proceedings of joint meeting of the Mining and Metalurgical Institute of Japan and the American Institute of Mining, Metallurgical and Petroleum Engineers,* September 1976, p. 287.

[12]LaMotte, R., 'Deepsea Ventures Pilot Run is Successful', *Ocean Industry,* October 1970, p. 7-9, 11, 13.

[13]*Barrons,* October 1976, p. 18.

[14]Anon., 'Converted Drillship will Begin Deepsea Mining Test in October', *Ocean Industry,* June 1977, p. 79-84.

[15]*Mining Magazine,* January 1973, p. 7.

[16]United Nations, Third Conference on the Law of the Sea, *Economic Implications of Seabed Mineral Development in the International Area,* Report of the Secretary-General, A/CONF.62/25, 22 May, 1974, p. 15.

[17]Frank, R.A., Testimony before subcommittee on Public Lands and Resources of the U.S. Senate Committee on Energy and Natural Resources and the Committee on Commerce, Science and Transportation, 19 September 1977.

[18]Wright, R.L., *Ocean Mining: An Economic Evaluation,* U.S. Ocean Mining Administration, U.S. Department of the Interior, May 1976, 18 p.

[19]Dugger, G.L., E.J. Francis and W.H. Avery, 'Technical and Economic Feasibility of Ocean Thermal Energy Conversion', *Joint Conference of ISES and Solar Energy Society of Canada,* Winnipeg, 15-20 August 1976.

[20]Szabo, J.L., U.S. Patent No. 3,983,017, 28 September 1976.

[21]Tinsley, C.R., 'Future Markets for Nodule Metals', *Proceeding of joint meeting of the Mining and Metallurgical Institute of Japan and the American Institute of Mining, Metallurgical and Petroleum Engineers,* September 1976, p. 249-266.

[22]Agarwal, J.C., N. Beecher, D.S. Davies et al., 'Processing Ocean Nodules: A Technical and Economic Review', presented at the American Institute of Mining, Metallurgical and Petroleum Engineers' Annual meeting, New York, 1975.

[23]Sridhar, R., 'Thermal Upgrading of Sea Nodules', *Journal of Metals,* December 1974, p. 18-22.

[24]Tinsley, C.R., 'Processing — No Longer a Problem', *Mining Engineering,* April 1975, p. 53-55.

[25]Brooks, P.T., and D.A. Martin, 'Processing Manganiferous Sea Nodules', U.S. Bureau of Mines, RI7473, January 1971, 19 p.

[26]United Nations Conference on Trade and Development, *Exploitation of the Mineral Resources of the Sea-bed beyond National Jurisdiction: Issues of International Commodity Policy, Case Study of Cobalt,* TD/B/449/Add.1, 26 June 1973, p. 11.

[27]Anon., 'Conceptual Evaluation of Deep Ocean Nodules', U.S. Bureau of Mines, Memorandum, February 1975.

[28]Meiser, H.J. and E. Muller, 'Manganese Nodules: A Further Resource to Meet Mineral Requirements?', Morgenstein, M. (Editor), *Papers on the Origin and Distribution of Manganese Nodules in the Pacific and Prospects for Exploration,* International symposium, Honolulu, Hawaii, 23-25 July 1973, p. 115.

[29]Montcrieff, A.G. and K.B. Smale-Adams, 'The Economics of First Generation Manganese Nodule Operations', *Mining Congress Journal,* December 1974, p. 46-50.

[30]Tinsley, C.R., 'Economics of Deep Ocean Resources — A Question of Manganese or No-Manganese', *Mining Engineering,* April 1975, p. 31-34.

[31]Victory, J.J., 'Mining Vessel to be Tested in 18,000 ft. Water in 1977', *Ocean Industry,* August 1976, p. 84.

[32]Graham, J.R. et al., *Process and Apparatus for Mining Deposits on the Sea Floor,* U.S. Patent No. 3,456,371, 22 July 1969.

[33]Ball, J., 'New Concept for Lifting Nodules', *Ocean Industry,* June 1967.

CHAPTER XI

THE FUTURE OUTLOOK FOR THE NODULE INDUSTRY

1. Introduction

During the past 15 years, the major effort in the development of deep ocean mining has been directed at extensive exploration at sea and the development of processing and mining (nodule collector) hardware on land. This exploration effort, which has been concentrated on the deposits of the North Central Pacific, leaves little question as to the potential value of nodule deposits. From a market point of view the principal metal to be obtained is nickel, followed by copper and cobalt. Limited amounts of manganese will also be produced by at least one consortium. On a smaller scale and under certain market and production conditions, manganese could be the principal revenue producer.

Extensive studies backed by data obtained thus far in the laboratory and in the field continue to show that the mining of manganese nodules is potentially attractive, provided that certain objectives can be met. Assuming that these objectives are attainable, success depends in part upon the outcome of large-scale tests now being planned or conducted. A paramount question is this: can the nodules be mined and brought to the surface reliably and economically, and in sufficient quantities to make a venture financially attractive? The minimum quantity is considered to be between 1 to 3 million dry metric tons per year. 'Dry' is stressed because the nodules contain about 30 percent sea water, and production numbers would otherwise be misleading.

Present economic studies indicate that approximately half of the cost of a nodule operation stems from the processing, while the remaining costs can be equally divided between mining and transporting the nodules to a shore processing plant. The large-scale tests will be limited to the surface platform, the 15,000 ft. of pipe and the bottom miner or dredge head.

There are four major consortia actively developing deep ocean mining systems: Ocean Mining Associates (OMA), Ocean Management Inc. (OMI), the Kennecott Group and the Ocean Minerals Company (OMC). Each of these consortia is comprised of companies from various countries, including the United States, United Kingdom, Japan, the Netherlands, Federal Republic of Germany Canada and Belgium. There is also the Group AFERNOD consisting of French companies. An international consortium was also formed by about 20 companies from six countries (France, Canada, Australia, Japan, USA and Federal Republic of Germany) to develop the continuous line bucket (CLB) system.

All of the major consortia have selected the same basic design for a mining system, i.e., a nodule collector on the ocean floor connected by a near vertical pipe to a surface ship. It appears that for most of the systems, pumping the nodules to the surface through the pipe will be accomplished by air lift. Nevertheless, an alternative approach using mechanical pumps will be tested by some of the consortia. Little is known of the status of the continuous line bucket system or the plans for testing it on a large scale and, therefore, it will not be discussed here.

1.1 Technical Risks

The technical risks associated with the development of a deep ocean mining system can be divided into five major categories and various subcategories. The five major categories are; bottom miner, lift system, surface system, transportation system and processing system. (Table 1).

TABLE 1.
Risk Assessment — Manganese Nodule Mining System

Bottom Miner	Lift System	Surface System	Transportation System	Processing System
Mobility	Pipe String Dynamics	Draw Works	Slurry Transfer	% Fines
Nodule Pick-Up	Slurry Lift	Derrick	Dynamic Positioning	Raw Materials
Nodule Transfer	Tool Joints	Pipe Handling	Tonnage	Process Time
Sediment Separation	Lift Pumps	Dynamic Positioning		% Recovery
Buffer Storage	Buffer Storage	Slurry Storage		Power
Reliability	Reliability	Slurry Transfer		Size
Control	Trip Time	Power		Product Quality
Power	Pipe Material	Control		Process Control

In the large-scale tests of the next two years, major emphasis will be given to the first three. One can subjectively identify varying degrees of technical risk associated with the functions and subsystems listed. For example, a high degree of reliability associated with mobility and nodule pick-up performance must be achieved so that confidence can be obtained in the designed production rates.

It will be noted that many of the items listed have a strong operational slant; that is, the success of the test relies as much on operational procedures previously established or acquired as it does on the design and technology incorporated into the system.

The list given is not complete but does contain many of the most important items. Some of the items are associated with efficient sediment separation and slurry lift. The bottleneck to production rates is primarily the slurry production of the lift pipe. Here sediment separation at the ocean floor is important because every ton of clay lifted reduces, in an equal amount, the total tonnage of nodules lifted. While tests and analysis onshore have provided reliable engineering data, only large-scale tests at sea can provide the accuracy desired.

The design of the mining system comprising the ship, the pipe and the nodule collector, draws upon technology developed in offshore drilling as well as upon government sponsored ocean research and development programmes such as the Glomar Challenger and Glomar Explorer. In the interest of economy, the ocean mining industry is using existing surface ships modified to handle the pipe and the nodule collector. Both Ocean Management Inc. (OMI) and Ocean Mining Associates (OMA) have scheduled their initial operations in 1977, while the Kennecott Group and Ocean Minerals Company (OMC) are looking toward 1978 and 1979 for their initial test operations. Some of the tests will be conducted in previously selected mine sites.

1.2 Mining Costs

All systems are being designed to mine at a slow speed of a few knots. The practical operating speed depends upon a number of parameters such as pipe stress and fatigue, hydrodynamic drag and stability as well as the total system dynamics. The dynamic characteristics of the nodule collector at the ocean floor water-clay interface, as well as the nodule collection and clay rejection efficiencies are also important elements to be considered. It is apparent that the tests will provide the necessary data for design modifications and improved operational procedures.

The cost of the feedstock to the processing plant is a very imporant factor in the profitability of a deep ocean mining system. Bulk shipping costs are probably the best known in the complete system. Next is the processing cost, based upon the data that have been obtained from the pilot plant work. Known with least accuracy are the mining costs. Here the large-scale tests at sea can provide a great deal of accuracy. Listed below are the important factors that determine the costs of nodule production.

Productivity — tons per day
Reliability — mean time between failures
Maintainability — downtime, time between overhaul
Costs — initial costs, operating costs
Operability — manpower, supporting equipment requirements
Complexity — number of parts, percent of high technology components
Controllability — miner maneuverability, nodule feed and flow

These are not independent variables. For example, reliability, maintainability and costs are interrelated as are many of the other factors.

Preliminary calculations show economy of scale. That is, 3 million tons per year may be produced with measurably less cost per ton than for one million tons per year. The number of crew required is almost independent of ship size as well as miner capacity. Energy requirements generally increase with size by a power of two-thirds. These types of scaling factors may not hold for a mining ship because it will be operated differently than other ships. Therefore, it is the operational characteristics that may have the greatest influence upon design.

All preliminary designs and prototype systems tested at sea exhibit characteristics that will require modifications. Not all the characteristics that lead to inefficient operations will be immediately recognized. For this reason, several trials at sea over a period of a year or so will be necessary. At each step, modifications will be made to provide better operational data for the next step. The cumulative knowledge gained will then be incorporated into the initial operational model.

1.3 Nodule Environment

It should be emphasized that manganese nodules represent somewhat of a unique ore deposit, a thin horizontal strata approximately 2 in. thick, extending for many miles. The nodules are friable, consist of pores (50 per cent by volume), and are imbedded in the bottom clays to approximately half of their vertical dimension. These clays are very soft and have some of the physical characteristics of heavy grease. The efficiency of the system is set not only by the percentage of the area that can be navigated by the nodule collector and the percentage of the nodules within that area that can be recovered, but also by the quantity of clay that is recovered along with the nodules. The pipe is sometimes a bottleneck for most efficient production; therefore, it should be operated close to the maximum design capacity.

There is considerable variation in nodule coverage as well as nodule ore grade, even when the area is extensively covered with nodules. Testing of the prototype systems by the different mining groups will most probably be accomplished in potential prime mine sites. Prior to the deep ocean deployment, shallow water tests (600 to 6,000 ft.) close to the West Coast will probably be conducted for initial checkout and crew training. Any major deficiency can then be more easily corrected because the mining ship is close to a supply and maintenance base.

Of particular interest is the area called Siliceous Ooze running from south of Hawaii to south of Baja California. These clays are somewhat unique since they are physically very weak. The shear strength of these clays varies from about 1/4 to 2 psi. The dredge head or miner must, therefore, be designed light enough to prevent sinking in these soft sediments. These soils are not frictional soils like those found on land, but instead are almost completely free of the fine grain-like sand particles of less than one micron, about half of which are composed of biological remains. The water content is very high—from one to three times the solid content. When disturbed, these soils weaken, going from a plastic to a thick liquid state.

1.4 Dredge Head

As approximately 90 per cent of the nodules at any location are on the ocean floor surface, the dredge head must be able to scrape the ocean floor within close tolerance over a width of 25 to 50 ft. or more. The dredge head must be carefully designed so that a minimum of clay is picked up with the nodules or, if too much is picked up, so that it is discarded prior to pumping to the surface.

There is a great variability in nodule size, abundance and ore grade as well as soil strength and ocean floor topography. The dredge head must be designed to provide a reasonable production rate and, at the same time, to navigate the ocean floor in a reasonably prescribed manner, avoiding obstacles and abrupt changes in terrain. Therefore, one of the primary functions of the mining system is accurate station keeping. One means for accomplishing this is, basically, the dynamic positioning system developed in offshore drilling; it is being used extensively by the offshore oil industry. A sonar transducer on the ship hull obtains distance from transponders several miles apart. The information is fed into a computer which then commands the ship propulsion and thruster systems allowing the ship to navigate accurately within a few feet of a prescribed course. It is important that the deviations of the dredge head from the ship's track are not excessive.

The second important capability of a mining system is the ability to lift heavy loads. The prototype systems require a capability of between 1 and 2 million lb., while the operational system will require a 2 to 4 million lb. capability or greater. Until recently, drill vessels and crane barges had capabilities of 1 million lb. or less. The Sedco 445 will have approximately 1.5 million lbs., while the Glomar Explorer has an 18 million lb. capability. Therefore, the state-of-the-art can provide all the pipe lift required. Another characteristic required is that the pipe system be gimbaled to allow the ship to pitch and roll while the pipe remains relatively vertical. This places less stress on the pipe, lengthening its usable life by limiting fatigue.

1.5 Efficiency of System

Mining efficiency estimates have been made by a number of consortia. Efficiency may be defined as the percentage of nodules by weight that is recovered from the total weight of the nodules available in any one mine site area.

The overall efficiency is made up of three elements — minable area, sweep efficiency and mining or dredge efficiency (see also Chapter VIII). The design of the total mining system will influence all three factors. The types of terrain that can be navigated by the dredge head will determine the amount of overlaps or gaps in the covered area. The percentage of nodules picked up by the dredge head on any one pass will determine the mining efficiency. Estimates are that the first systems will pick up about a quarter of the available nodules and that, in time and with improvments, one could expect the efficiency to approach 40 to 50 per cent. The initial large-scale tests now planned by the industry will provide the productivity and cost figures required to determine the cost of the feedstock to the processing plant.

1.6 Legal Environment

Large-scale tests of deep ocean mining systems will be conducted during 1977-1979. 'Large-scale' means from one-fifth to full designed production rates. The expenditures during this period of time for each consortium can vary from a low of $20 to $30 million to a high of $75 million or more. Based upon the performance of their respective mining system designs, each industry group will be in a position to make decisions on the next step. The costs involved can vary from $500 million to $1 billion and include the processing system and its infrastructure as well as the transportation and full scale mining systems. However, it should be stated that the mining industry is carefully watching the United Nations Law of the Sea negotiations as well as United States legislative prospects and will speed up or slow down its investment pace as warranted by variables of the legal environment. The minimum investment by American industry in the next 5 to 7 years could be approximately $2 billion. It should be emphasized that the technology is there; the large-scale tests will provide the engineering, operational and economic confidence. Hopefully, there will be a favourable legal environment to provide a favourable investment climate, thus allowing this potentially very promising enterprise to be established.

2. Ocean Mining Associates (OMA)

This group is composed of U.S. Steel, Union Miniere of Belgium and the Sun Co., Inc., with Deepsea Ventures under contract as project manager. The group has essentially completed its tests of the dredge head on land in a simulated sea floor test pit. Previously, in 1970, Deepsea Ventures demonstrated a nodule mining system on the Blake Plateau off the Florida coast in about 3,000 ft. of water. Although the nodule and sea-floor characteristics are different than in the Pacific nodule belt, the test proved the ability to collect nodules and air lift them to the surface. OMA has conducted extensive exploration of the nodule area and, in particular, of the mine claim site that the company has announced. This has been accomplished with the R/V Prospector, a well-equipped oceanographic vessel.

Current plans involve the conversion of an ocean ore carrier (the Weser Ore, a diesel-driven 560-ft. vessel) for the mining test platform. The dredge head is an ocean bottom screening and collecting device connected to a near vertical pipe. The system is towed through the water at a speed of from one to three knots. A 27-ft wide by 34-ft long moon pool, or center well, is installed amidship to handle the pipe and dredge head. Thrusters fore and aft will provide the precision navigational capabilities required and long, base-line acoustic navigation will be employed. The pipe will be handled hydraulically from a gimbaled platform. Other modifications include pipe and dredge handling equipment aboard ship, air compressors for the nodule air lift system and cargo holds and distribution systems for handling the nodules. Quarters will be provided for an additional 30 people who will handle the complete mining system.

The system tested is approximately one-fifth scale. Initial deep-sea tests began in November-December 1977. The group has not announced planned operations beyond these initial tests or the time scale of the full production system.

3. Ocean Management Inc. (OMI)

The OMI group consists of AMR (Metallgesellschaft, Preussag, Rheinbraun, Salzgitten), Federal Republic of Germany, Domco-Sumitomo companies and other Japanese firms; International Nickel Co., Canada; and SEDCO Inc., U.S. The basic philosophy of the group is the maximum use of the technology of the participants to accomplish as much "in-house" as is practical. The project is managed by Ocean Management Inc. of Bellevue, Washington.

The exploration programme is being carried out by the research vessel Valdivia, an excellently equipped ocean exploration vessel. It is a German vessel equipped by the Government of the Federal Republic of Germany and staffed by a German crew. The scientific crew is drawn from the participants' staffs and rotated as appropriate according to the skills needed. The prime tools are free fall samplers with cameras, box corers and underwater TV. The research vessel Valdivia has a narrow beam depth recorder, low frequency profiler, satellite navigation, chemical analyses facilities and seismic equipment. Data reduction can be accomplished aboard, except for a few computer-based operations.

The large-scale tests at sea will be conducted with an ocean floor nodule collector towed from the long vertical pipe on the ship. Various collector designs have been tested in a local facility built for the purpose. Initial tests at sea with the collector head minus the pipe have been made. The Sedco 445 was chosen for the large-scale test.

The deep-sea tests, scheduled for November-December 1977, are reportedly underway. The primary pumping method will utilize submerged hydraulic pumps with a 9⅝ inch pipe string reaching to full depth; air lift will also be tested. The Sedco 445 was converted to use a special hydraulic pipe lifting system with heave compensation, positioned on a gimbaled platform. The ship has a dynamic positioning system employing thrusters both fore and aft. The ship is 445 ft. long, with a beam of 70 ft. and a moon pool of 37 ft. diameter.

4. The Kennecott Group

The Kennecott group consists of Noranda Mines of Canada, Rio Tinto Zinc, British Petroleum and Consolidated Gold Fields of United Kingdom, Mitsubishi of Japan and Kennecott Copper of the U.S. As is the case of the other groups, the consortium has attempted to bring together as great a capability as possible to provide the knowhow required for the venture.

Kennecott has, in the past, devoted considerable effort to exploration; and it is believed to have more than one potential mine site available, with one of undisclosed location selected for the initial operations. This group has leased well-equipped vessels for its exploration efforts.

Kennecott appears to have completed its basic development work on its nodule collector and is now planning its large-scale test at sea in 1979. The previous efforts involved a test mining unit of approximately 2,000 ton per day capacity. It was tested at a 15,000 ft. depth, but no nodules were brought to the surface. The planned operations will involve the towing of the dredge head at the end of a near-vertical pipe string from the surface ship. It appears the previous tests made with the dredge head were accomplished by towing with a cable. Evidently this is the technique used by other groups for initial tests. Apparently, the design of both the miner and the lift system is fixed. Kennecott has not announced the ship to be used for the tests.

5. Ocean Minerals Company (OMC)

Lockheed has formed a consortium with Amoco Minerals Co. of Standard Oil of Indiana, Billiton International Metlas of Royal Dutch Shell and Bos Kalis Westminster. Both Billiton and Bos Kalis are located in Holland. Lockheed's activities in ocean mining have extended over a period of more than 14 years and include the development of mining and processing technology.

Lockheed has developed a test miner and has essentially completed the required land tests in a simulated sea floor "mud pit" as well as in a wet tank. The company has not yet selected a ship for the at-sea tests; therefore, it will probably be sometime in 1978 that these initial tests will be made. Consideration has been given to the Glomar

Explorer as well as to drilling vessels. It is not known at this time if the Glomar Explorer will be available when required. If not, or for other reasons not optimal for the test, another ship will be used. The test miner has a production capacity of approximately 20 to 25 per cent of the planned operational unit. The successful tests of this miner will provide engineering data required for the production miner.

6. The French Group

AFERNOD (Association Francaise pour l'Etude et la Recherche des Nodules) was set up in 1974 between CNEXO (Center National pour l'Exploitation des Oceans), CEA (Commissariat à l'Energie Atomique) and SNL (Le Nickel—Société). Later, Chantiers de France Dunkerque (from the Empain Schneider Group—and BRGM (Bureau de Recherches Géologiques et Minieres) joined the group. The initial main objectives of the French Group were to undertake studies in order to determine the feasibility of nodule exploitation. Extensive exploration works have been carried out and mining studies have been conducted. A mini-pilot processing plant was built and work has begun in it. Furthermore, economic evaluations are also being undertaken. In the next stage, the French Group intends to expand its work regarding mining problems; in particular it will undertake sea tests within the areas identified during the exploration stage.

PART V

THE PROCESSING PROBLEMS

CHAPTER XII

PROBLEM ADDRESSED AND A SUMMARY OF THE DISCUSSION

1. Problem Addressed

The efforts expended to overcome some of the technical problems described in the papers and discussions summarized in the previous chapters would be scientifically interesting but commercially of no significance if the valuable metals contained in nodules could not be recovered. Three or four metals, namely nickel, copper, cobalt and, in some cases, manganese have to be recovered in some commercially acceptable form. A quick examination of grades, however, indicates that even in those deposits classified at the meeting as potential reserves, the first three metals account for less than 3 percent of the total weight of a nodule. Between 25 and 30 percent by weight of the nodule is made up of manganese and the remaining gangue consists of several clay materials including kaolinate, phillipsite and chlorite, together with iron, silica and water. Certain properties of nodules contribute to the further exacerbation of the problem: (i) Nickel, copper and cobalt are not found as separate minerals in nodules. Instead, they are distributed in a matrix of manganese oxide. (ii) It follows that like most laterite and garnierite deposits, and in contrast to sulphide deposits, nodules cannot be beneficiated by low cost physical means. (iii) Nodules are extremely porous, so much so that their moisture content is about one-third of their weight.

The complex composition of nodules presents one of the most difficult problems because it confronts any metallurgical process with impurities of almost every element in the periodic chart.

Agarwal presented a concise report on the problems involved in recovering metals from nodules and the way in which they might be overcome. The facts clearly pointed to nodules as being essentially a nickel ore: copper and perhaps cobalt would be recovered as coproducts, while manganese would have second priority. The very heavy energy consumption and therefore cost of either carrying or evaporating more water than was absolutely essential in the recovery process were emphasized. These considerations together with others had led the Kennecott Group to adopt a completely new process that involved the reduction of the manganese dioxide at atmospheric temperatures and pressures by leaching with ammonia, followed by liquid ion extraction and conventional electrolytic recovery of copper, nickel and cobalt. The recognition of nodules as an important new resource had justified a major research programme aimed at optimizing their use.

2. Summary of the Discussion

One of the economic problems associated with processing nodules was caused by their high water content. After the transport of ore containing 30-40% water to the processing plant, it would then be too costly to expend considerable energy evaporating the moisture. The Kennecott group had made a major breakthrough in metallurgical chemistry to solve this problem; they found a way to reduce the metallic oxide matrix under ambient temperature and pressure.

Differences in capital costs of various processes arose largely from the different equipment used in the various processes, since both pyrometallurgical and hydrometallurgical routes were material handling systems. Drying the ore prior to treatment was considered to be too costly because of energy consumption, and was ruled out by the Kennecott group, hence the pyrometallurgical route was eliminated. Some leaching processes were considered unacceptable because they also recovered many impurities, such as bismuth and strontium, which posed quality problems for the nickel, copper and cobalt products. In regard to the choice of a reagent, Agarwal mentioned the difficulties in finding a compliant reagent which did not vigorously attack the manganese matrix. The option of producing manganese was left open in the Kennecott process, since it was possible that market conditions might improve, although Agarwal emphasized that nodule-derived manganese could not really compete on existing markets with high-grade manganese ore from land.

Concerning market impacts of metal production from nodules, Agarwal suggested that metal production from nodules could, in the medium-term, expand the markets for certain metals. The high price of cobalt, for example, was related to its limited availability rather than to production costs. It could be much more widely used in alloys, but its use was discouraged by its price. Production from nodules and the penetration of cobalt into new markets would stimulate its use in alloying and could expand the total market.

CHAPTER XIII

COMPARATIVE ECONOMICS OF RECOVERY OF METALS FROM OCEAN NODULES

1. Background

Ocean nodules represent one of the world's largest abundances of nonferrous metal values in existence today. The contained nickel, copper, cobalt and manganese are present in enormous quantities, enough to supply hundreds of years of demand at current consumption levels. Nevertheless, ocean nodules cannot be expected to displace current metal resources until economically competitive methods for the mining and the processing of this ore are developed. The required technology for mining from the ocean bottom, transporting and processing the nodules is expensive, complex and sophisticated.

Processing of ocean nodules is a key concern. Capital investment for the process plant generally overshadows that for mining by about a factor of two. Even if the substantial technological challenge of bringing these nodules from the ocean bottom to the surface is met, nodules will not be an attractive source of nonferrous metals until economical techniques for extracting the metal values are available.

Because the higher grades of ocean nodules are found to occur 1000-1500 miles from the nearest landfall, the cost of transporting material containing 30 to 40 percent moisture is significant. The nature and potential of the nodule deposit has been well delineated,[1 2] however, a brief review of those properties most important to the selection of an extraction process will be useful. Like most oxidized deposits and in contrast to the sulfide deposits, nodules cannot be easily beneficiated by low cost physical means. Nodules usually contain about 30 percent by weight of a gangue fraction which consists of several clay minerals including montmorillonite, chlorite, kaolinite, and phillipsite, and biogenic fractions of calcium carbonate and silica. No separate minerals of copper, nickel and cobalt have been found in nodules so that the concept of their liberation size has no application. Instead, the valuable metals are distributed in the manganese oxide phases: todorokite, birnessite, and MnO . Early claims were made that cobalt was present in the iron phases,[3] ferric hydroxide gels and goethite, but it now appears that the majority of cobalt in ocean bottom nodules is in a manganese phase.

Nodules are friable and easy to grind with a Bond index for closed circuit rod milling of about 7 kilowatt hours per ton.

Nodules contain extremely fine pores of the order of 100 A° diameter and this results in porosites of nearly 60 percent by volume and surface areas of about 200 m²g.[4] The nodule pore structure has led to consideration of their use as an adsorbent and catalyst.[5;6,7,8,9,10,11] As a result of this porosity, the moisture content of nodules is about 30-40 percent by weight. Many nodule extraction schemes involve direct leaching because the cost of removing the water is a problem for all processes involving drying, high temperature reduction, or pyrometallurgy.

It might be assumed that the porosity of the ore would make direct leaching rapid.

However, Han has provided data showing that several days are necessary for direct acid leaching at low temperature.[12] As a result, several of the processes under commercial consideration use reducing agents to attach the tetravalent manganese oxide matrix, simultaneously releasing the valuable metals, eliminating the slow leaching problem and making manganese available for recovery if desired. Potential reductants available as bulk commodities are hydrogen chloride, ferrous ion, sulfur dioxide, and carbon.

The complex composition of nodules presents at once one of the great potentials of nodules. Any metallurgical process is confronted with impurities of almost every element in the periodic table.

Ocean nodules contain about 1-2 percent nickel, 1-2 percent copper, 0.1-0.5 percent cobalt, and 25-35 percent manganese on a dry basis. The mined nodules size consistency will vary depending on the mining method used. The delivered material may vary from a finely comminuted sludge to whole nodule chunks. Any nodule processing plant must be able to deal with the problems of material handling as well as the presence of substantial sea water.

2. Characteristics of Nodules Processing

All nodule processes basically involve reduction of manganese in the nodules to liberate the metals, separation of liberated metals from the manganese and gangue, separation and refining of these metals to yield copper, nickel and cobalt products. These processing steps can be carried out in two basic ways—hydrometallurgically or pyrometallurgically.

The metallurgist faced with the problem of developing a process for nodules will initially review the contained metal values (Table 1). The first reaction is that this is a contaminated manganese ore. However, a consideration of the market force shows that this wet, impure, 30 percent ore cannot compete economically with dry, pure, 45-50 percent manganese ore available at low cost commercially. It is not likely that manganese from ocean nodules will be competitive with high-grade terrestrial ores as long as they remain commerically available. The stockpiling of the manganese-containing tailings probably makes good sense as a hedge against a worldwide disruption in manganese ore supplies. Because Pacific Ocean nodules are of higher grade than those from the Atlantic Ocean, any U.S. operation producing manganese from nodules will be on the Pacific Ocean and therefore badly located with respect to the largest U.S. steel makers.

TABLE 1.
Metal Values per Dry Short Ton of Ore

	Grade	Amount	Price	Value
Manganese	30.0%	600 lb.	$ .175*	$105
Nickel	1.3%	26 lb.	2.10	55
Copper	1.1%	22 lb.	0.60	13
Cobalt	0.2%	4 lb.	6.00	24
				$197

The next highest value is nickel. Nodules contain as much or more nickel as many commercial laterite deposits. Although the mining cost will be higher, some of the other metal values, notably copper, can be recovered to pay for mining.

The decision to develop and use a process that produces, or does not produce, a manganese product is largely one involving the demand for manganese as compared to nickel. Figure 1, which is based upon United Nations' statistical data shows that the growth of nickel production in the world over the 1966-1975 period was 6.3 percent per year which is in excess of the 5.6 percent growth rate in the World Industrial Production Index. On the other hand, world manganese output has grown at only 3.8 percent

FIGURE 1.

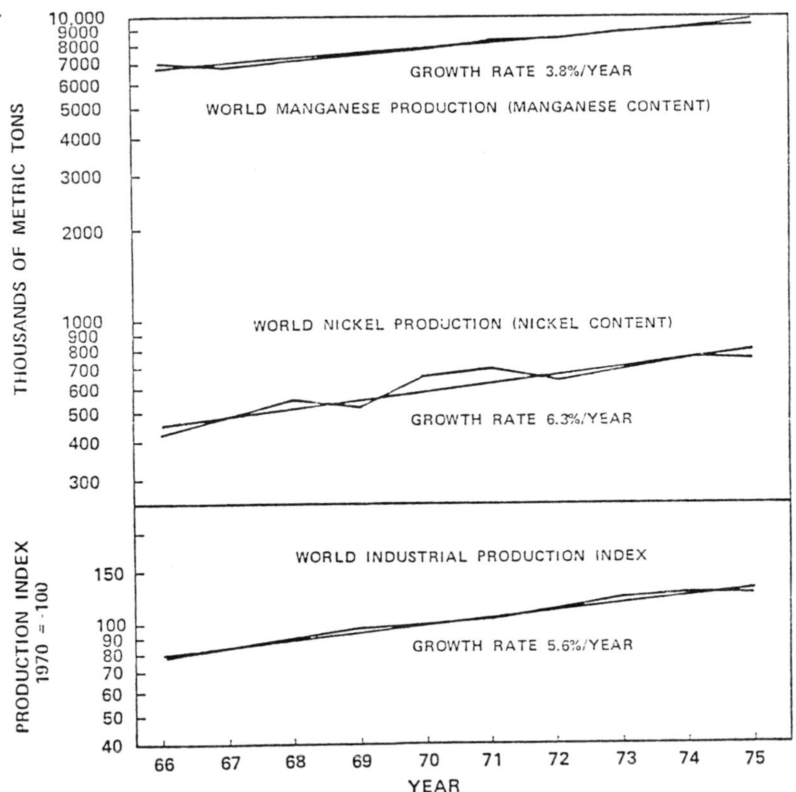

Source: United Nations Statistical Yearbook

per year over the same period. A decision to commit substantial capital funds into the facilities to produce manganese from ocean nodules is an extremely difficult one to justify when growth rates are substantially less than the growth in industrial production. If long-term market studies show that nickel growth rates will continue to outperform the industrial production index, then it is likely that a new producer could expect to be able to move his nickel output into the world market without a serious deterioration in price.

Nickel as a prime recovery target has much to recommend it. The demand is growing at 6-7 percent annually. High-grade ore supplies, unlike those containing manganese, are not as abundant. The price has fluctuated, but in general has been steadily rising. Cost effective substitutes for most nickel applications do not exist.

Accepting nickel and copper as the principal products, one must select a process route. Smelting might be considered first. Note, however, that the ore contains 30-40 percent moisture and is not amenable to physical benefication. The entire ore must, therefore, be dried and raised to smelting temperature to recover less than 3 percent of its weight as metal values.

But in smelting we will be producing a manganese slag that could be used to produce ferromanganese and, thus, increase revenue. Unfortunately, this slag will be far from ideal for this purpose because it will be fairly low grade (35-40 percent) and will contain many impurities, especially copper and phosphorus. Even if this manganese slag is assumed to be delivered at zero cost, it probably cannot compete with terrestrial ores at today's prices. (There is, of course, some price at which it can compete.)

A further consideration is that the commercial venture becomes dominantly a manganese business. Assuming 90 percent recovery, the manganese will produce $95 revenue per ton of ore and nickel, copper and cobalt $83. This locks the venture into competition with a highly integrated manganese industry. Capitalization to sales ratio will be high.

Furthermore, at the large scale of operations necessary for an economic ocean mining operation, the manganese and indeed the cobalt mined will be a significant portion of the market (Table 2).

TABLE 2.
Comparison of 5,000 tpd (1,500,000 typ) Nodule Mine Output with Metal Markets

	Mining Rate (short tons)[1]	Percent of U.S. Use[2]	Percent of World Production[3]
Nickel	19,500	9.1%	2.1%
Copper	16,500	0.8%	0.2%
Cobalt	3,000	48.4%	7.7%
Manganese (all forms)	450,000	34.4%	4.1%

[1]Contained-assuming 100% metal recovery. Nodule analysis: 1.3% NI, 1.1% Cu, .2% Co, 30% Mn, 10% Fe.
[2]USBM-1976-Total U.S. Industrial Demand
[3]USBM-1976-World Mine Production

It thus appears that any nodules commerical venture will be more risky if it is locked into the necessity of manganese production to make a profit.

A nodule processing venture that is primarily oriented to nickel must be competitive with the nickel operations coming on line that are processing laterites.

In the hydrometallurgical nodule processes, the metals are solubilized from the nodules by a strong leachant or by a combination reduction-leach process. The leaching can be done by one of the common mineral acids such as sulfuric acid or hydrochloric acid, or by a base such as caustic or ammonia. The preferred base is ammonia because it can be recovered and recycled in the process.

The hydrometallurgical processes most commonly proposed are:

Solution reduction and ammoniacal leach
High temperature reduction and ammoniacal leach
High pressure sulfuric acid leach
Low pressure hydrochloric acid reduction leach.

On an overall basis, there is considerable flow sheet similarity in the front end material handling of nodule ore and in tailings disposal. There are also similarities in the methods used for recovery of metals from solution and also in the final production of copper, nickel and cobalt products. The major differences are found in the leaching system (including pretreatments to improve leachability) and in the method of recycling the reagents.

Sulfuric acid is the best known and cheapest of all hydrometallurgical reagents. It has been tried with nodules. However, according to the work of both Han and Hoover,[13] leaching at ambient temperature and pressure requires several days, results in solubilization of substantial amounts of manganese and iron and reaction with clay gangue and consumes as much as half a ton or acid per ton of nodules. As a result of considerable work, it has been found that a high pressure leach, 500 psig at 450 °C, can extract nickel, copper, and cobalt with negligible solubilization of manganese and iron and with acid consumption of about 0.3 ton per ton of nodules.[14] [15] [16]

This identifies the preferred route for sulfuric acid leaching, but it introduces the high cost of pressure equipment for a corrosive environment. The cost of even 600 lbs. of sulfuric acid per ton of nodules is substantial, and this acid is not economically recyclable.

One is, thus, impelled to look for another recyclable reagent. Hydrochloric acid is a possibility but it has major disadvantages.[17] The acid in reacting with nodules solubilizes all the metals, including manganese and iron. The manganese is reduced so that part of the chloride ends up as chlorine, often a difficult by-product to sell, or expensive to convert back to hydrochloric acid. Much of the chloride is tied up with manganese, as $MnCl_2$, and with the other metals. Recycling this chloride is difficult. We have not found in hydrochloric acid a suitable inexpensively recyclable, selective reagent.

This reasoning suggests that a reduction-ammoniacal leach process as developed by Caron and used first at Nicaro for nickel laterites should be advantageous.[18] [19] The nodule ore is easily reduced. The ammonia-ammonium carbonate leach selectively extracts nickel, copper and cobalt as the ammine complexes and leaves iron and manganese in the tailings. There is little or no reaction with the clay gangue and the reagent is recoverable by simple distillation. If the economics works for laterites, it should work for nodules, with copper and cobalt paying the mining cost and additional profit. Finally, we will have in our tailings pile a low-grade manganese ore than can be used if the price of competitive ores rises to make this economically justifiable.

A hydrometallurgical process that does not have manganese recovery as its primary objective is the solution reduction-ammoniacal leach (Cuprion) process.

Reduction is carried out directly on a slurry of nodules at low temperature and pressure using carbon monoxide as the reducing agent. The aqueous phase reduction eliminates the need for water removal and drying which are costly and troublesome components of the high temperature reduction step.

The other major category of process is the pyrometallurgical process. Typically, it involves drying the nodules, prereducing the nodules in rotary kilns, electric furnace smelting followed by blowing in converters to produce nickel-copper matte. The matter is leached in sulfuric acid and the metals recovered from solution.

A general flow sheet comparison of the pyrometallurgical approach with the Cuprion hydrometallurgical approach is given in Figure 2.

3. Cuprion Hydrometallurgical Process

In the low temperature ammoniacal leach process, the wet nodules are delivered to a pond for storage. Nodules, dredged from the pond, are sent to a ball mill for wet grinding to a fine mesh. Grinding is required to assure adequate kinetics during reduction. Grinding requirements are relatively moderate. The work index is about half of that for copper ore. It is also expected that a considerable fraction of the nodules will be substantially comminuted prior to reaching the process due to the mining operation.

Reduction is carried out on the solid slurry in agitated tanks at low temperature and low presure. The reducing agent used is carbon monoxide which may be generated by partial oxidation of fuel oil.

The copper, nickel and cobalt values which are partially released by the reduction operation are further solubilized in a countercurrent decantation (CCD) wash carried out in a series of thickeners. Leaving the overall CCD wash is a pregnant solution containing the copper, nickel and cobalt metal values and a washed tailing which contains essentially all of the manganese as manganese carbonate. The ammonia and carbon dioxide reagents are stripped from the tailings and recycled in the process prior to tailings disposal. The metals are extracted from the loaded pregnant solution. Liquid ion exchange is used to remove copper and nickel and sulfide precipitation is used to remove cobalt.

4. Pyrometallurgical Process

In the pyrometallurgical process ore is stored in a stockpile rather than ponded since sea water content should be kept to a minimum. Grinding is not necessary. The moisture content of the nodules is an important consideration. It is expected that the nodules may contain about 45 percent sea water. The nodules are dried in a rotary dryer using pulverized coal as fuel. The fuel gas from the dryer is passed through an electrostatic precipitator to remove entrained solids and then to a stack for dispersion into the atmosphere. No special SO_2 removal equipment is used since the nodules are assumed to have absorptive capability for SO_2.

The nodules are prereduced using coal in a rotary kiln prior to smelting in an electric furnace. The manganese reports to the slag. The matte from the smelter is transferred to the converter where a matte containing about 25 percent Cu, 35 percent Ni, and 5 percent Co is produced.

The matte represents about 4 percent of the weight of the original nodule feed on a dry basis. This matte is assumed to be hydrometallurgically treated for extraction of metal values.

The metals in the matte are solubilized using oxygen pressure leaching. Copper,

FIGURE 2.

HYDROMETALLURGICAL PROCESS
NODULES
STORAGE POND
GRINDING
NH_3 CO_2
CO
CO GENERATION
FUEL OIL
WASH
C.C.D.
STRIPPING
STEAM
LIX
Ni E.W.
Cu E.W.
H_2S
Co RECOVERY
TAILINGS
ORE PRETREATMENT
CUPRION REDUCTION
LEACHING & REFINING
PYROMETALLURGICAL PROCESS
NODULES
STOCKPILING
COAL
DRYING
COAL
KILN REDUCTION
SLAG
OFF GAS
SLAG
ELECTRIC FURNACES
CONVERTERS
MATTE
LEACH
C.C.D.
LIX
Ni E.W.
Cu E.W.
H_2S
Co RECOVERY
SLAG
ORE PRETREATMENT
PYRO REDUCTION
LEACHING & REFINING
AUXILIARIES

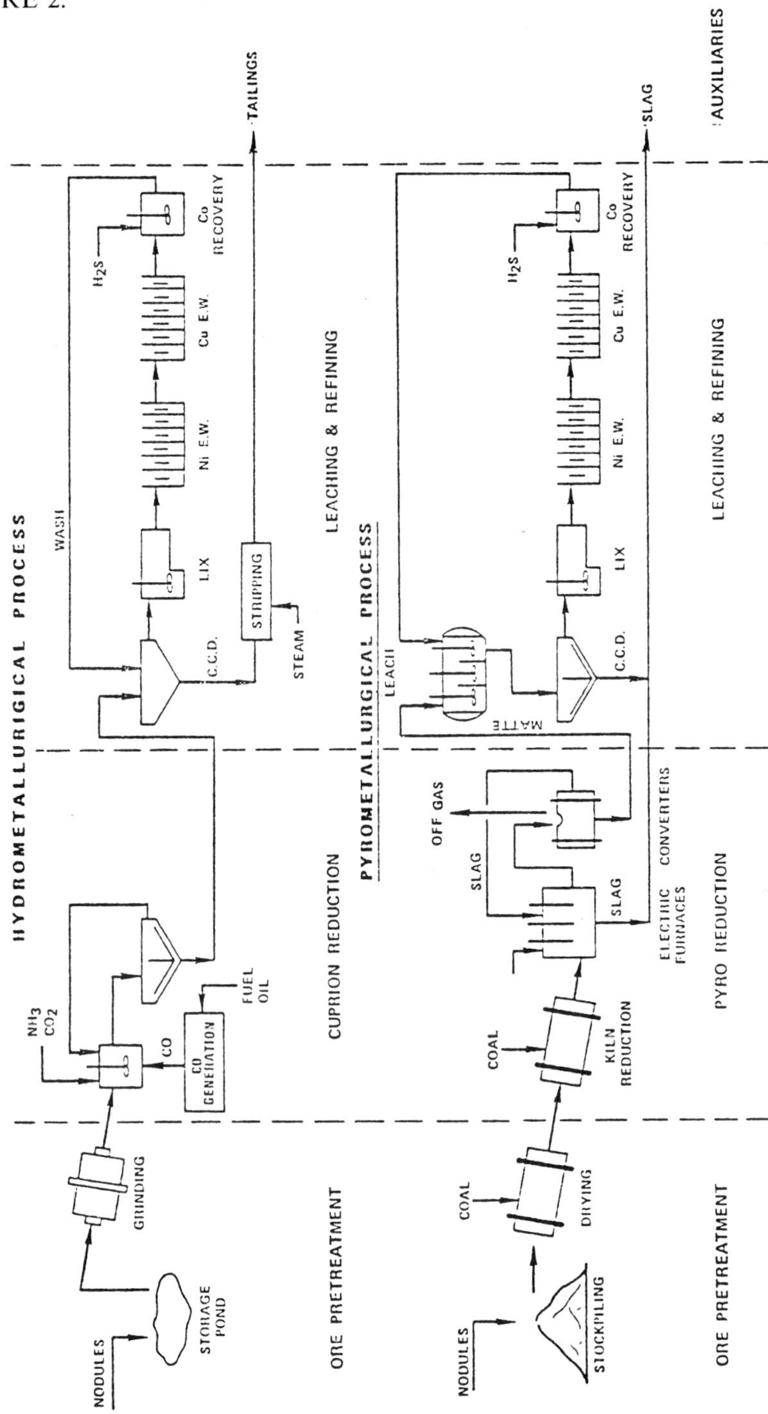

nickel and cobalt are separated and recovered using liquid ion exchange and electrowinning in a similar manner to that used in the hydrometallurgical-based processes. The liquid flow dependent equipment items are considerably smaller, however, since pregnant liquid flow rate is lower by a factor of 6. The costs for nickel electrowinning, copper electrowinning and cobalt recovery are functions mainly of the quantity of metals produced and the costs associated with these items tend to predominate.

5. Technical Aspects

Overall nickel recovery is about 90% from each process. Copper recovery in the hydrometallurgical process is also about 90 percent. Our analysis indicates that the smelter slag (pyrometallurgical process) should contain over 0.2 percent copper, which corresponds to an overall copper recovery of about 80 percent. The lower copper recovery by the pyrometallurgical route, however, is compensated by a higher cobalt recovery. A higher recovery of molybdenum is also probably possible by the pyrometallurgical process. However, a slag cleaning furnace is necessary if this is to be accomplished.

Both the hydrometallurgical and the pyrometallurgical processes produce a high manganese-containing waste material, namely tailings, in the case of the hydrometallurgical process. These can be processed to a synthetic manganese ore or still further to ferromanganese and silicomanganese.

6. Economic Analysis

Each process can be subdivided into four categories:

Ore pretreatment
Reduction
Leaching and refining
Auxiliary operations

In the case of the pyrometallurgical process, the flow sheet includes the recovery and reuse of as much of the waste heat as possible. the electric energy was also conserved by prereduction of the nodules in kilns using coal.

The capital and operating cost profile for the hymetallurgical and pyrometallurgical process are shown in Table 3. The capital cost for the pyrometallurgical process runs almost 60 percent higher than that for the hydrometallurgical process. Operating cost, excluding depreciation, is about 40 percent higher. The primary reasons for the higher capital and operating costs for pyrometallurgical process are ore drying and high temperature reduction of the manganese, as compared to the lower temperature and in-slurry reduction of manganese without need for ore drying in the hydrometallurgical process.

7. Manganese Production Option

Table 4 compares the capital and cash operating cost profiles for producing copper, nickel, cobalt and manganese products from nodules by the hydrometallurgical versus pyrometallurgical process. To obtain an unpurified manganese ore of similar quality to that produced by the pyro process as a slag increases the capital investment of the hydro process by roughly 25 percent. Both processes require further purification of the unpurified manganese ore in order to produce a product which might be suitable for sale. Additional capital and operating costs required for manganese recovery from Cuprion process tailings brings the two comparative routes close together, so the process choice is dictated by location and market strategy.

Manganese ore produced from nodules could be still further upgraded to form ferromanganese and ferrosilicon products. The costs for doing so would be the same for each process.

TABLE 3.
Hydrometallurgical Process versus Pyrometallurgical Process Cost Profile

	Pyrometallurgical		Hydrometallurgical	
	Capital Cost	Operating Cash Cost	Capital Cost	Operating Cash Cost
	($ MM)	($ MM/yr)	($ MM)	($ MM/yr)
Ore Preparation	40-65	5-10	25-35	3-7
Ore Drying	40-80	15-25	——	——
Reduction	150-250	50-70	50-70	25-35
Leaching & Metals Prod	100-150	25-35	150-200	25-35
Auxiliaries	40-60	15-25	40-60	25-35
Total	450-550	100-150		
Total	450-550	100-150	250-350	70-120

TABLE 4
Cost Summary: Cuprion Process versus Pyrometallurgical Process Profile

	Pyrometallurgical		Hydrometallurgical	
Products	Capital Cost Percentages	Oper. Cost Percentages	Capital Cost Percentages	Oper. Cost Percentages
Cu, Ni, Co	60%	49%	43%	29%
Mn Ore Unpurified	——	——	11	10
Mn Ore Purification	4	6	5	7
Cu, Ni, Co, Mn Ore	64	55	59	46
FeMn and SiMn	36	45	41	54
FeMn and SiMn	36	45	41	54
Cu, Ni, Co, FeMn, SiMn	100%	100%	100%	100%

Figure 3 shows the effect of sale of manganese on both the overall project profitability and investment requirements for both processes. Production and sale of manganese has a very large positive impact on the profitability of the pyrometallurgical pro-

FIGURE 3.

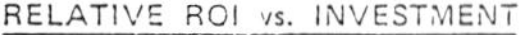

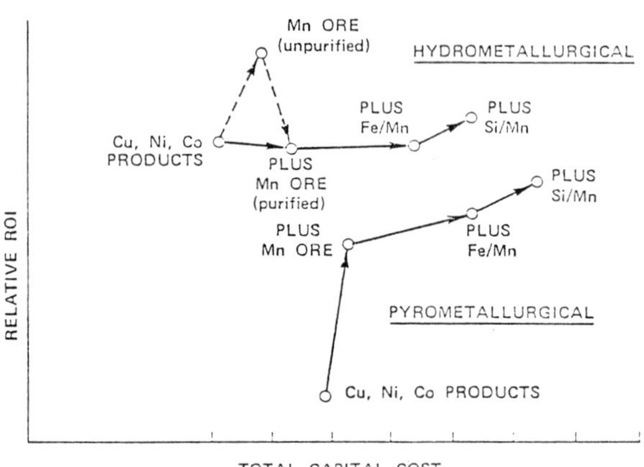

cesses while production and sale of manganese has only a minimal impact on the hydrometallurgical process. The reason for this is that the slag produced in the pyrometallurgical process is already a dry impure manganese ore, which only requires a low cost purification step to yield the synthetic manganese ore. In the case of the hydro process, the tailings slurry containing manganese carbonate must be upgraded by flotation drying and calcining before going to the purification step.

As contrasted to the pyrometallurgical process, the Cuprion hydrometallurgical process allows decoupling the production of copper, nickel, and cobalt from the production of manganese without adverse impact on the economics allowing for a lower initial investment.

8. Conclusions

It is apparent from examining the world production statistics for nickel that nodule-derived metal will be principally competing for new nickel markets and will not supplant existing production. Manganese production strategy is more complex and indicate, from the business decision viewpoint, that manganese should only be produced if world demand cannot be met from terrestrial ores. It is apparent that not all consortia interested in nodules will produce manganese as priority product, and that any added manganese capacity will only be able to compete for new markets.

Notes and References

[1]Cronan, D.S., 'Authigenic Minerals in Deep-Sea Sediments', *The Sea,* Goldberg, E.D. (Editor), Vol. V, Wiley Interscience, New York, 1974, p. 491-526.

[2]*Ferromanganese Deposits on the Ocean Floor,* Horn, D.R. (Editor), National Science Foundation, Washington, 1972, 293 p.

[3]McCutchen, H.L. and P.H. Cardwell, 'Process for the Electrolytic Refining of Heavy Metals', U.S. Patent 3,855,089.

[4]Fuerstenau, D.W., A.P. Herring and M.P. Hoover, 'Characterization and Extraction of Metals from Sea Floor Manganese Nodules', *Transactions of the Society of Mining Engineers,* Vol. 254, 1973, p. 205-211.

[5]Bartlett, R.W., 'Sulfation Kinetics in Sulfur Dioxide Absorption from Stack Gases', Report by Stanford University, Process Metallurgy Group, 1972, 157 p.

[6]Chang, C.D. and A.J. Silvestri, 'Manganese Nodules as Demetalation Catalysts', Industrial and Engineering Chemistry Process Design and Development, 13(3), 1974, p. 315-316.

[7]Skarbo, R.R., 'Extraction of Copper and Nickel from Manganese Nodules', 1973, U.S. Patent 3,723,095.

[8]Weisz, P.B., Yardley and A.J. Silvestry, 'Demetalation of Hydrocarbon Charge Stocks with Manganese Nodule Catalyst', 1974, U.S. Patent 3,813,331, 10 col.

[9]Weisz, P.B., et al., 'Demetalation of Hydrocarbon Change Stocks', 1973, U.S. Patent 3,716,479, 24 col.

[10]Weisz, P.B., et al., 'Demetalation of Hydrocarbon Change Stocks', 1973, U.S. Patent, 3,716,479, 24 col.

[11]Wilder, T.C., 'Two-stage Selective Leaching of Copper and Nickel from Complex Ore', 1973, U.S. Patent 3,736,125, 6 col.

[12]Han, K.N., *Geochemistry and Extraction of Metals from Ocean Floor Manganese Nodules,* Ph.D. thesis University of California/Berkeley (Unpublished), 1971, 212p.

[13]Hoover, M.P., *Studies on the Dissolution of Cu, Ni, and Co from Oceanic Manganese Nodules,* M.Sc. thesis University of California (Unpublished), 1966.

[14]Hanig, G., 'High Pressure Sulfuric Acid Leaching of Manganese Nodules', *Second International Conference and Exhibition for Ocean Engineering and Marine Sciences, Proceedings,* Dusseldorf, 1973, p. 432-444.

[15]Hubred, G.L., *An Extractive Metallurgy Study of Deep-Sea Manganese Nodules with Special Emphasis on the Sulfuric Acid Autoclave Leach,* Ph.D. thesis, University of California/Berkeley (Unpublished), 1973, 220 p.

[16]Ulrich, K.H., U. Scheffler and M.J. Meixher, 'Processing of Manganese Nodules by Acid Leaching', *Second International Conference and Exhibition for Ocean Engineering and Marine Sciences, Proceedings, Dusseldorf, 1973, p. 445-457.*

[17]*Cardwell, P.H., 'Extractive Metallurgy of Ocean Nodules', Mining Congress Journal,* November 1973, p. 38-43.

[18]Boldt, J.R., Jr., *The Winning of Nickel,* Methuen and Co. Ltd., London, 1967, 487 p.

[19]Brooks, P.T. and D.A. Martin, 'Processing Manganiferous Sea Nodules', U.S. Bureau of Mines Rept. of Invest. 7473, 1971, 19 p.

PART VI
OTHER TECHNOLOGICAL ISSUES

CHAPTER XIV

PROBLEM ADDRESSED AND A SUMMARY OF THE DISCUSSION

1. Problem Addressed

The metallurgical technology for the processing of manganese nodules and the economic factors involved in the choice of a particular processing route were studied in the previous chapter. This chapter elaborates further on technological issues that arise in selecting sites for manganese nodule processing plants. In the location of processing plants, the traditional considerations concerning the proximity to the mines or to metal markets lose their significance in the case of deep-sea manganese nodules. While, as a general rule, a coastal area will be preferable, such factors as supply of energy and environmental considerations will play a dominant role in the selection of a specific site.

In view of the absence of past experience and lack of precise estimates in the feasibility studies, a ranking of all the relevant factors in order of importance is not possible. Therefore, a division of these factors into primary and secondary groups may be adopted. The former, which includes energy supply, environmental considerations, harbour approach and handling facilities and investment climate, is sufficient to eliminate certain areas as candidate sites. The latter, which includes size of the area, infrastructure, personnel and materials, serves as a second level screening mechanism for those sites which satisfy the primary factors. Since all the deposits likely to be exploited in the near future are in the Northeast Pacific region, the study by Diederich evaluated sites within the circum-Pacific region, notably the west coast of the United States. The prospects for locating processing plants in the developing countries within this region were also considered.

The issue of developing countries taking part in the technological development of deep-sea mining is attracting attention, not only in the traditional context of technology transfer, but also from the point of view of the participation of all nations in harnessing the deep-sea resources that have been characterized as the 'common heritage of mankind. The paper by Kuchen discussed transfer of marine technology. Transfer in this context went beyond transfer of technological information, and even beyond acquisition of technology, to the application and further development of technology. Cooperation was deemed a prerequisite for the transfer of deep-sea technology, and given the relative infancy of the research and development effort in ocean mining technology, it was felt that it offered special opportunities for cooperation between industrialized countries and developing countries. Emphasizing the importance of the participation of developing countries in the research and development effort at this "early" stage, the paper explored effective ways and means of facilitating their participation.

2. Summary of the Discussion

After the presentation of the papers, the discussion among the Group of Experts focused on a number of distinct critical aspects in the selection of sites for nodule processing plants, namely, the regional location of the plant or plants, the energy requirements of the process selected and its implications, the lead times required to take account of environmental considerations and the relative merits of locating such plants in developing countries.

Following the pattern set in the study, the discussions also focused on attractive sites within the circum-Pacific region. It became apparent that in the absence of a specific metallurgical process, the number of "attractive sites" satisfying the many factors, including harbour approach, availability of power and investment climate, could be potentially great in number. With the specification of an optimum metallurgical process implying a hierarchy within the factors, however, the number of such potential sites would be substantially reduced.

On a matter related to location, the experts felt that a shipboard or offshore location for the process plant was impractical given the fact that nodules would be delivered to the plant in the form of a slurry. A shipboard operation would require a substantial amount of storage space which would introduce a significant cost item into the economics of any venture.

Considerable time was devoted by the Group to the problem of estimating the power requirements and identifying the source of this power. Nuclear, hydroelectric and coal as sources of energy were discussed in this context. Even though it was recognized that metallurgical processes would be energy intensive, it was felt that the nature of the energy source was dependent on the energy profile of the process and was therefore process specific. With reference to the specific sources of the power, nuclear energy was deemed the least attractive, particularly if it constituted the sole source of power. It was suggested that later generations of mining technology might employ this source of power, but the matter was not discussed in any depth. Hydroelectric power and coal were considered very attractive, and coal in particular, being readily available on the west coast of the United States, was viewed by some experts as being most suitable for the purpose.

Regarding environmental considerations, the requirements of local, state and Federal authorities were touched upon. In an illustration of this point, reference was made to the problems that confronted a chemical company that wanted to locate a plant on the west coast of the United States. It was emphasized that considerable time had to be allotted to the approval procedures associated with locating plants in certain industrial countries. In terms of a time horizon, it was also stated that this activity should be undertaken within the planning and design stage which would take about five years to complete. The actual construction of the plant, it was suggested, would take approximately two years.

Concerning the relative merits of locating processing facilities in developing countries, a question was raised as to the attractiveness of a site in the Gulf of Mexico. In the opinion of one of the experts, the distances to be traversed to reach any site within this region might place too great an added burden on the economics of the total venture. In any case, a complete evaluation would require a detailed analysis which had not been carried out. In general, regarding the suitability of developing countries as sites for processing facilities, the following were cited as possible advantages: the availability of energy and/or the potential to expand existing sources; more flexible environmental requirements; and the possible catalytic effect on the development process. An issue of concern, however, related to security of investment in these countries. Overall, however, it was decided that this was an area for cooperation and negotiation among the parties concerned.

Following the presentation of papers on technological aspects of deep-sea mining, the Group of Experts took up the question of the means for transferring sea-bed mining technology to developing countries. Recognizing that this technology was capital intensive, one expert suggested that developing countries could participate initially in the sampling and surveying activities required in general prospecting and improving the data base, then by training a few technicians and equipping some of their commercial vessels with fairly inexpensive gear. Other experts were of the opinion that this might not really serve any purpose since developing countries would be preoccupied with other more pressing problems. It was agreed that despite their importance, these issues were best discussed in other forums.

The Group agreed with the following conclusions based on the discussion:

(1) A coastal location for nodule processing plants was preferable to a shipboard, offshore or inland location. Given the location of deposits of current interest, the first processing plants would be limited to the circum-Pacific region, perhaps the west coast of North America.

(ii) Specification of a particular metallurgical process would influence the selection of a site for the processing plant.

(iii) Energy supply was one of the most important factors in determining the actual site. The selection of an energy source was dependent on the selection of a process route; although for the first set of plants, the use of coal appeared most likely.

(iv) Environmental considerations were important determinants in site selection and the lead time required to meet particular environmental concerns would vary from place to place.

(v) In the circumstances that a developing country could provide, the materials required and then the investment climate in the country would be among the major factors that determined whether a site in that country was selected.

(vi) Transfer of ocean mining technology could be meaningful only when developing countries participated actively and went through the process of 'learning by doing'.

(vii) Transfer of ocean mining technology might involve different considerations from those in traditional transfer of technology given the fact that it was still in the research and development stage.

(viii) The specific form of cooperation among the industrialized and developing countries would have to be worked out in negotiations in other forums.

CHAPTER XV

SPECIAL ASPECTS OF SITE SELECTION FOR MANGANESE NODULE PROCESSING PLANTS

1. Introduction

Some introductory remarks are called for before entering into a discussion of site selection for a manganese nodule processing plant. Considering the present status of ocean mining, and site selection as part of it, we must admit that we are entering into an area where technology has not yet been developed and does not lend itself easily to experiences and approaches of other fields. Analytical evaluation of practical experiences concerning the location of large projects is not available in scientific publications. Nor has there been a study on a scientific approach covering the spectrum of elements pertinent to decisions and procedures for selecting a site. In view of this state of affairs, it is necessary to evolve a practical method for the site selection of a nodule processing facility.

It becomes increasingly evident that new determining factors have to be taken into account, for which new scales of priorities have to be chosen. These priorities are determined mainly by the shifting global economic situation in which the availability and use of energy play an important role, by the intensity of industrialization, and by the limits of the general growth rate. For many countries, conservation of energy and maintenance of the quality of the environment are reflected in national legislation and are important factors which must be taken into account in any industrial development. Consequently, today's site selection procedure must give additional attention to socio-economic and demographic criteria.

2. Site Selection as Part of a Feasibility Study

The main issue involved is not to find the best one amongst a multitude of possible sites. The basic question is whether the project can be realized, and if so, what kind of difficulties might arise. A site selection study undertaken as part of a feasibility study thus should encompass details of prices and expenses and the assessment of possible impact and probable trends.

A further and essential aspect consists of deciding between the static assessment of site realities (i.e. values at a given time) and the dynamic representation (i.e. extrapolation of static values to the future possible date of investment). The latter does not apply to essential site factors in those cases where scientifically or practically deductable parameters cannot be identified with sufficient accuracy.

It is very important to bear in mind tendencies and possibilities of development in certain relevant fields, such as power supply and environmental control. In this context, a forecast of probabilities will show the necessary consequences.

3. Procedure of Localization

Since most of the prospecting and exploration cruises have been concentrated in the region southeast of Hawaii, it would seem appropriate to presume that a mine-site could be located somewhere in that region. In the light of the need for cost reduction and for avoiding or reducing negative factors, the potential site for processing is likely to be limited to the circum-Pacific coasts. These coastal areas, by no means, should be regarded as offering the same economic-geographic and socio-economic facilities, since the circum-Pacific region covers countries with different economic backgrounds and at different stages of development. On the basis of characteristics of (i) geographic and ethnical uniformity, (ii) socio-economic level of development, (iii) techno-economic infrastructure, and (iv) investment climate and stability, the circum-Pacific region can be subdivided into six sub-groups which may be considered as potential areas for site selection.

Starting with the criterion of the distance from the mine site to the processing plant, the American coastal region (Nos. 5 and 6, Figure 1) appears attractive. The distance criterion is certainly of great significance, since it involves a significant part of the financial load in terms of operating and investment costs through the whole operating period of the project. The greater the distance, the longer it requires to make a round trip, and thus the greater the expenses. This is particularly true for ships of a size between 50 to 70 thousand tons.

Within the two chosen regions a further delimitation can be made by applying the following basic requirements:

(i) a harbour or a pier must exist or it must be possible to build one at a reasonable cost;

(ii) the approach through a shipping channel must be possible for ships with drafts between 40 to 50 ft. depth;

(iii) The technological and economic infrastructure is available or can be planned, or its enlargement is possible at a later stage.

(iv) large industrial projects are permissible from the standpoint of the landscape.

(v) For practical reasons, the location of a site should be limited to a distance up to 15 miles from the coastline, unless there is a harbour situated at the river.

When these basic requirements are applied, only a number of areas will be found suitable.

4. Studies on the American West Coast

In this section some figures will be given concerning requirements for a four metal processing plant with a capacity of 3 million dry metric tons per year. The results of the studies and their evaluation as far as the American west coast is concerned will be described.

Concerning the problem of weighing the factors for determining a site, the criteria given above for selecting the site seem to be of equal weight and importance. However, on a closer inspection, they differ in many respects. To establish a set of site selection factors according to priorities and then classify the areas accordingly cannot be technically realized and logically sustained.

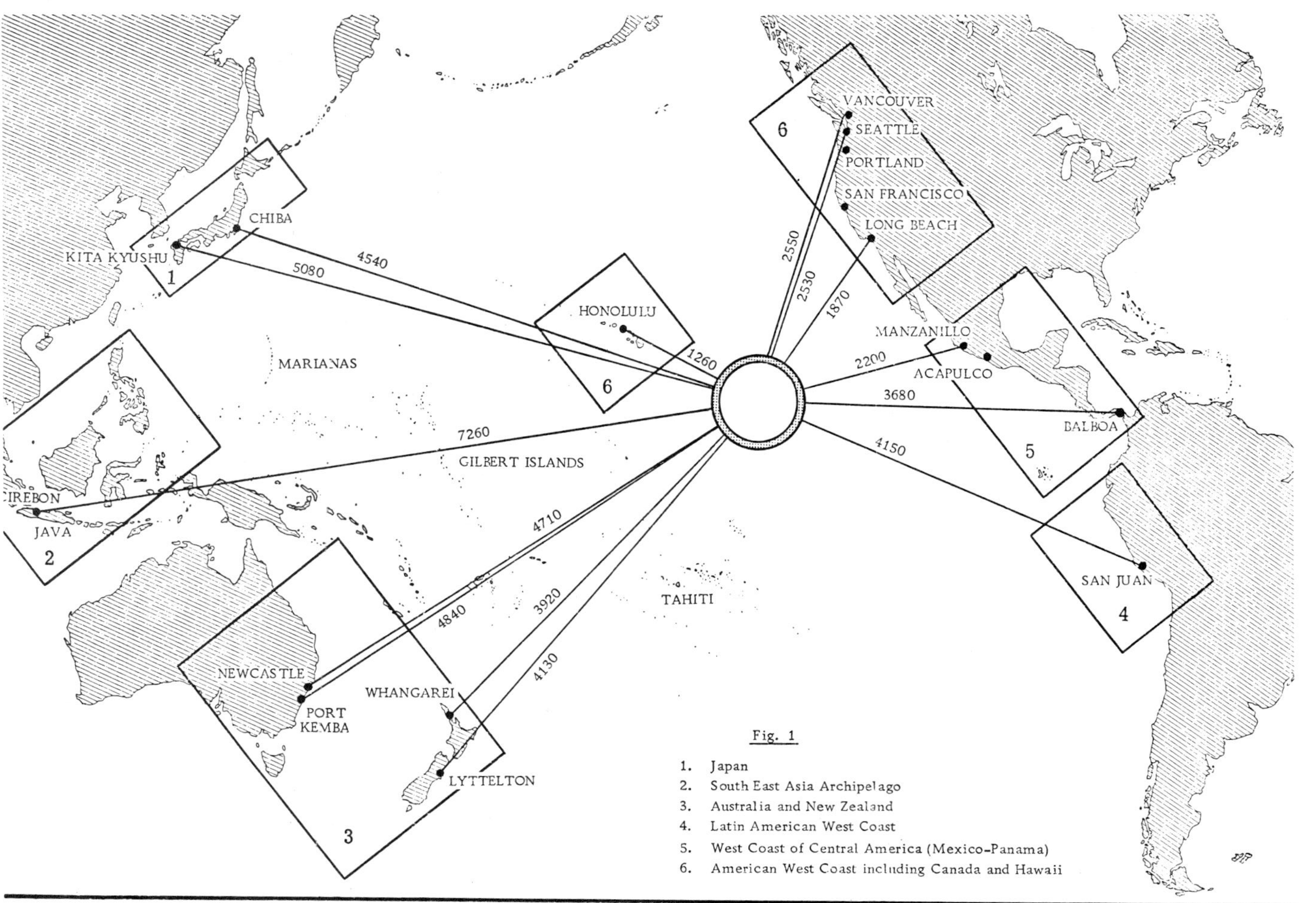

Fig. 1

1. Japan
2. South East Asia Archipelago
3. Australia and New Zealand
4. Latin American West Coast
5. West Coast of Central America (Mexico–Panama)
6. American West Coast including Canada and Hawaii

An approach suitable for practical application is to divide site selection factors into what we might call primary and secondary factors.

The *primary factors* are those project-specific requirements which will cause the elimination of the project if they cannot be fulfilled. In other words, if the project requirements do not come up to the site specific factors or cannot be realized at reasonable cost, the plant cannot be erected in the region in question. These factors include:

power supply
environmental control
approach to the harbour
investment climate.

The *secondary factors* comprise those requirements characterized by flexibility and adjustability. The conformity of these requirements with site-specific realities can be attained by organizational and financially justifiable measures. They include:

labor market (manpower situation)
infrastructure
supply of materials
the available land area

4.1 Size of the Area Needed

One of the most important factors for the size of a site is the "conception of the project": an approach of "centralization" i.e. unloading facilities, processing plant and waste disposal forming one local unit, or an approach of "decentralization", i.e. the above mentioned facilities are separated from each other.

If the first approach is accepted, the area needed will amount to 1.5 to 2.5 square kilometres including an area of 0.5 to 1.0 square kilometres for waste disposal. Apart from the size and topography of the site, its original use, ownership, and zoning requirements are also factors to be taken into consideration.

The investigations led to the following conclusions:

(i) all areas meeting the requirements had been either already earmarked for industrial uses or, presumably under the pressure for further development, had already been allocated for various uses.

(ii) a large part of each area is at present fallow or natural land and preparatory work (e.g. clearing and levelling) must be undertaken.

(iii) Some areas judged to be very suitable with regard to their logistics (near a river or ocean) require extensive preparation at high cost (e.g. due to their sandy or muddy bottom which require heavy foundations and piling).

The investigations show that the basic problem is the implications of the project and is not the availability of areas or the costs of the land (at about $1,000 to $10,000 per square acre, the amount of the total investment is marginal). Even after approval of its use as a processing site is granted, approval from other authorities dealing with various aspects of land use at different levels (e.g. county, state, Federal government) is still required. Neither the time to be spent on these matters nor the achievable result can be calculated beforehand.

4.2 Infrastructural Facilities

The infrastructural technical facilities are important for judging the approaches to the site from the open sea. These facilities include the pier, unloading facilities for high-sea carriers, installations for storage as well as appropriate devices to transport the nodules to the plant.

Investigations showed that:

either,

(i) most of the harbours do not have those facilities;

(ii) if such kinds of facilities are available, they are already fully utilized; they might therefore get overloaded thus causing break-downs;

or,

(iii) the facilities are available but a suitable area for the plant is missing.

As a consequence, new facilities specifically installed for the project are necessary.

The connection of the investigated sites to road and rail can be achieved everywhere; the costs for additional connection vary greatly. Generally, it can be stated that areas with a large harbour within an urban center do not offer favourable conditions. On the contrary, it is those areas with small and medium sized harbours which should be preferred.

4.3 Manpower

The number of persons required to operate such a plant can be assumed to range from 1,000 to 1,500 persons distributed about evenly into three groups: skilled, semi-skilled and unskilled. The level of wages, weekly working hours, working weeks per year, and the power and structure of trade unions are influential factors. On the basis of the labour market, note should be taken of the following findings and experiences:

(i) within a range of 20 to 30 miles around the investigated localities the number of inhabitants should be at least 30,000 to 50,000 in order to provide a labour force of 1,000 to 1,5000.

(ii) some relief of the labour situation may be expected because of the relatively high mobility of manpower. It is unnecessary therefore to limit the search for a suitable location to regions where heavy industries already exist.

(iii) wages vary between 5 and 8 dollars per hour. From these figures, a total amount of approximately $15 to 20 million per year with fringe-benefits of approximately 25 percent can be estimated. Wages in highly industrialized regions exceed those of less industrialized areas by about 30 percent at the present rate. This difference in wages may not continue for long since the influence of trade unions in the latter region are supposed to reach the same level within a rather short period of time.

(iv) Erecting the plant in a region distant from cultural centres and having climatic difficulties would considerably hinder the recruitment of manpower. The difference may range from 50 to 70 percent. These circumstances could not only endanger the safety of the plant operation, but also incur additional expenses on wages. The relatively high rate of unemployment amounting regionally from 6 to 10 percent does not allow the conclusion that a policy aimed at the creation of work would ease environmental requirements. The supply of manpower will not cause great problems if the outlined circumstances are taken into account.

4.4 Operating Materials

At the present time the supply of operating materials can be regarded as secured in all areas investigated. The supply of bunker-C-oil must be particularly watched as it amounts to a considerable part of the operating costs (about \$20 to \$25 million per year).

Depending on the processes of the plant, the requirements per year are:

60,000 to 300,000 tons of bunker-C-oil,
0 to 300,000 tons of coal and
several hundred thousand tons of various chemicals.

Between oil and coal, some substitution is possible. In view of the use of crude oil within the United States, it is evident that in the long run, other sources of power are needed or alternative processes must be applied which are not dependent on fossil fuel primary energy.

The requirements for water are quite high: i.e. 30,000 to 120,000 cubic metres of cooling water and 10,000 to 20,000 cubic metres of processing water per day. Attention should be given to the purity of the water and the temperature increase of the water to be discharged. The supply of cooling or processing water from existing networks is almost impossible in all locations. Independent water supply connected with the necessary treatment plant would be needed.

Although drawing water from rivers or lakes does not present any technical difficulties, the resulting environmental impact can hardly be assessed. Additional expenses and time are required for obtaining licences and for meeting other requirements. Locations situated in arid zones, or in areas which are poor in water supply, or are too distant from natural water resources will present problems of adequate water supply.

4.5 Climate of Investment

When evaluating the climate of investment, the foremost considerations are: the analysis of the over-all economic situation; the readiness to accept the newcomer; and promotional tools such as fiscal incentives, financial assistance and participation.

Within the regions under study a policy towards decentralization was noted, although with varying degrees of intensity. A further growth and expansion in areas already crowded with industries should be avoided. It is preferable to move into areas of small and economically weak industrial centres.

Some conclusions may be drawn with respect to the qualifications of various sites and the possible reduction of expenses:

(i) relief from high expenses connected with project-specific and infrastructural investments can be secured only by starting detailed negotiation with the appropriate authorities at an early stage.

(ii) no analogous conclusions can be drawn with regard to smelters, so far as taxation is concerned. An individual assessment as part of the promotional programme must be expected.

(iii) savings become feasible if support means are taken advantage of in connection with the establishment of secondary production plants. Consequently, a project which is based on the erection of a smelter alone, and without further processes to increase the values to be located nearby, has scarcely a chance of support and realization.

4.6 Environmental Control

The question of environmental control comprises three essential aspects: air pollution, contamination and heating of water, and waste disposal. The environmental impact of an industrial complex will be closely and critically evaluated on the basis of regulations, requirements and standards established at three different levels: Federal government, state government, and local authorities. The starting point is the so-called "Environmental Impact Study" which has to be undertaken by the applicant who wants to operate the plant.

Air pollution is measured by the atmospheric content of various pollutants and poisonous substances. The standards set for areas of heavy industrial concentration are so high that additional emissions will not be allowed in those regions. Since the operation of a processing plant can result in considerable pollution of the air, its location is an area of industrial concentration is no longer feasible.

Contamination or heating of rivers by waste and cooling waters is a serious problem. Strict measures are particularly enforced in regions which the socio-economic structure is related to fishing activities. In these regions, water treatment and systems for reducing the temperature of the discharged water are definitely required. Provision for disposal in close proximity to a plant which is situated near or at a river or in the coastal area, is practically excluded since the soil and the groundwater may be affected. Waste disposal facilities have to be located in arid and uninhabited areas. This means decentralization of disposal, a measure which will influence the economics of the project very considerably due to the mass of material to be transported and the cost of handling, unloading and storage.

From the standpoint of environment and its quality, the process to be applied in a plant is more important than consideration of costs. The procedure for obtaining a licence from the authorities concerned with environmental control might easily last from 4 to 6 years or even longer. This time requirement must be taken into account in the planning of ocean mining. During this period, it appears that no action on the plant can be taken since the result of the licencing process cannot be foreseen. For example, in January, 1977 Dow Chemical Co. suspended its plan to build a $500 million petrochemical complex in Solano County, in northern California, because it underestimated the time and cost of California's licencing procedure. It had to obtain 65 permits, 5 from the Federal government, 40 from the State of California and 20 from the three countries. In 1976 the company was only able to obtain four of these after having spent about $2 million per year and after a two-year fight for the project.

The requirements to maintain the landscape and to avoid infringing upon recreation provide another reason for close scrutiny of the expectations of the population, which has become quite sensitive to such kinds of threats. It is equally difficult to imagine that in a thinly populated area where the natural landscape is undisturbed, a large industrial complex could be erected free from any opposition. Thus, a great number of locations must be eliminated as possible sites. The site selection will have to concentrate on areas which are not densely industrialized and yet possess a certain amount of economic and operational activity.

4.7 Power

The demand for electrical energy is considerably high. The installation should provide 300 to 400 MW so as to supply an average consumption between 7 and 10 million kilowatt/hours per day. When evaluating the supply situation one must consider not only supply possibilities from power plants of hydro, fossil and nuclear types, but also the possibility of getting it integrated into a power grid.

The future demand shows an enormous increase, which will grow by about 100% within the next 15 to 20 years (Figure 2). Since hydro-resources cannot be further increased in the United States the basic load would have to be taken over in the 80's by coal-fired or by nuclear power plants, whilst hydro-power would be made available at peak load periods. Construction of thermal power plants not only requires very high investments, but is tedious due to the long, drawn-out licencing procedure through numerous authorities with a yet uncertain result. Thus, long-range planning to secure the power supply becomes very risky indeed.

An additional requirement amounting 300 to 400 MW cannot be fulfilled anywhere in the American west coast. Most probably it might not become available in the foreseeable future without long-range preparations and planning. This state of affairs is aggravated by the demand for an uninterrupted supply source. According to official estimates, a period of 10 to 15 years is required from the time of deciding on the erection of a power plant to the point of starting its operation and its going on-line (Fig. 3).

In order to obtain a reliable electrical power supply, the processing plant has to be integrated into the supply system at an early stage - at least 8 to 12 years before the power demand will actually arise.

The chance of a safe energy supply can, however, be improved by acquiring shares of an electrical power plant or by participating in its financing, an investment which can easily reach an amount of $500 to $600 million for a load of 300 to 400 MW.

4.8 Conclusions The following conclusions may be drawn regarding the west coast of

the United States: In principle it is possible to locate the project there. The need to secure a power supply for the plant's operational life, i.e. a range from 20 to 25 years, is a critical one. The complicated and time-consuming procedures for getting licences from various authorities concerned with environmental protection may cause delays rendering the planning process quite hazardous. Moreover, the outcome remains uncertain. The feasibility of the project should be judged from the standpoint not only of the time required for its realisation and the capital investment needed, but also of the volume of investment for project-accompanying measures, which are normally only a peripheral cost in conventional industrial projects. In this case, the establishment of an individual power station for operating the project needs to be considered. The cost can be high.

5. Considerations Concerning Location of Processing Plant in Developing Countries

Additional factors concerning the location of a processing plant in developing coun-

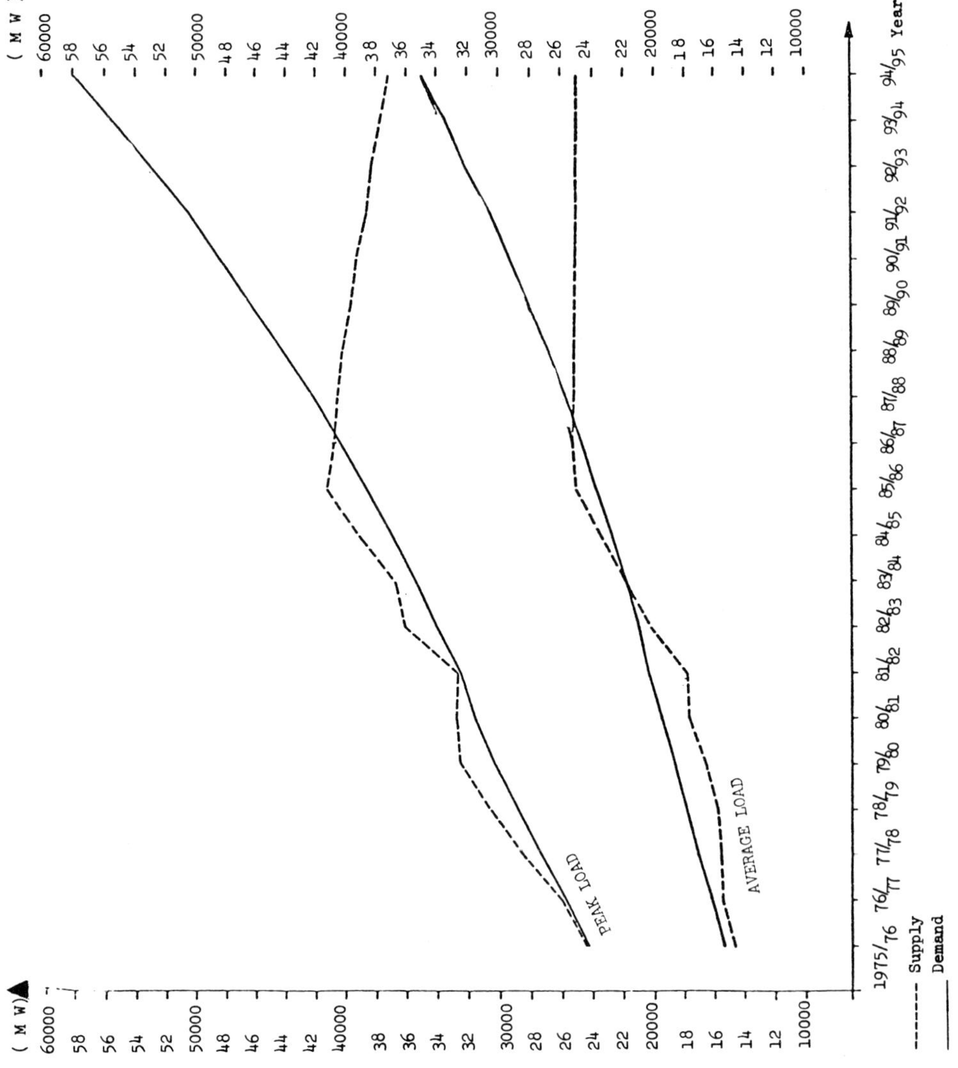

Figure 2. Energy Requirements.

TYPICAL NUCLEAR POWER PLANT CONSTRUCTION LEAD TIMES
Lead Time (years)
10
9
8
7
6
5
4
3
2
1
0
In-Service Date
Construction Time, Oregon
Oregon State Application
Construction Permit Application
Construction Permit Granted
Site Inves-tigations
State Review + PSAR Prep
State and Federal Review
Construction
Test
Begin Site Certification
Request for Bids
NSSS Award
Oregon State Approval Granted
Construction Time, Wash.
Request for Bids
Construction Permit Application
Construction Permit Granted
Site Inves-tigations
PSAR Prep
State and Federal Review
Construction
Test
Begin Site Certification
NSSS Award
Wash.State Application
Wash.State Approval
Hypothetical 6 Year Schedule
Federal + State Applications
Construction Permit Granted
State + Fed. Review
Pre Site Work
Construction
Test
NSSS Award
PSAR Submittal

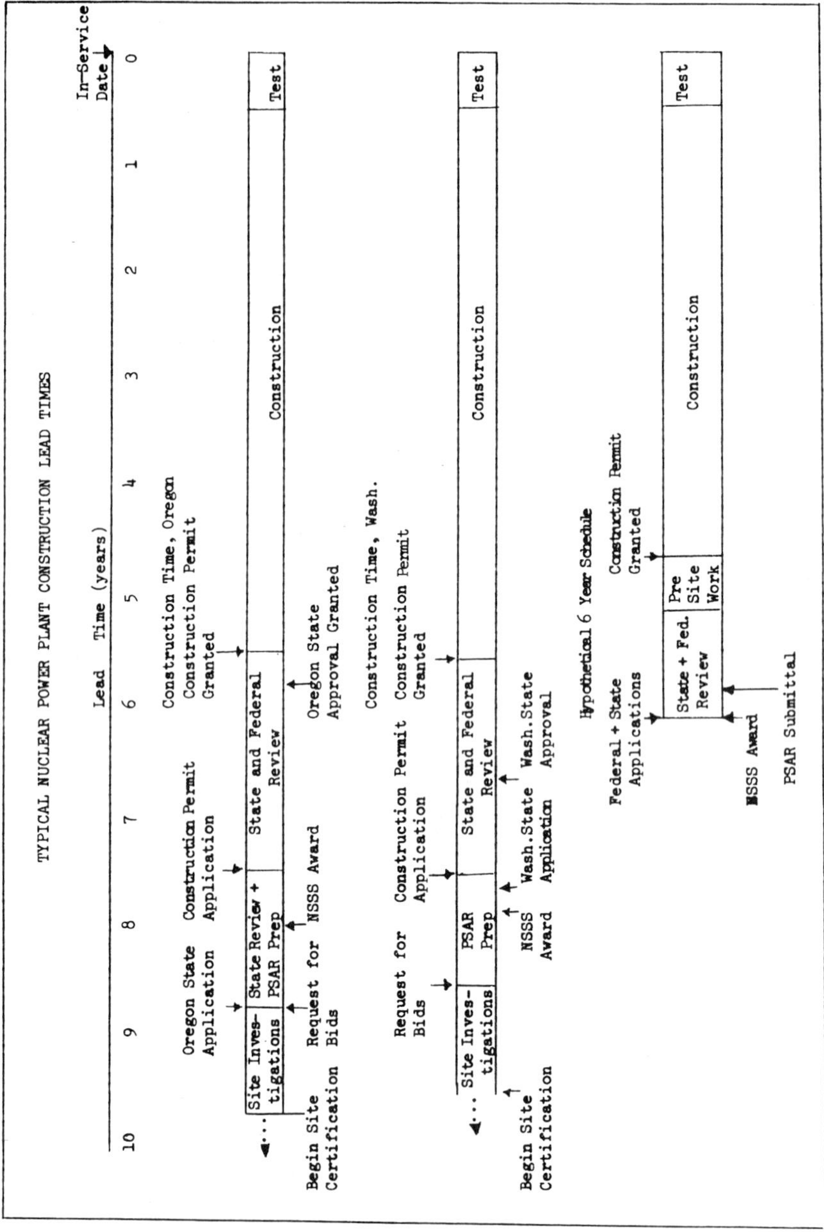

Figure 3

tries must be considered. Countries bordering the circum-Pacific coast might be preferred since they offer potentially well suited sites. The advantages might even be enhanced by comparatively low transport costs, which is true particularly for Latin American countries.

The above mentioned criteria for the selection of sites should be scrutinised carefully in applying them to developing countries since they could assume different degrees of importance and influence: new dimensions can be expected from transferring the frame of reference to developing countries.

Amongst the secondary criteria for selecting a site, the factor of labour market assumes a different perspective due to the high rate of umemployment, about 30 to 50%. The supply of unskilled labour at least is assured. In addition, manpower costs should be lower and constitute an advantage. However, skilled labour may be in short supply even if the generally considerable mobility is taken into account. The difficulties seem to be in training and education, and in importing operating materials in addition to investment goods are spare parts. These problems sometimes are rather difficult. Infrastructure, particularly that of transport and communications normally does not come up to the standards; power supply and the necessary networks together with the required reliability may also pose problems.

Amongst the primary factors, the climate for investments and its assessment assume the highest importance. Almost all developing countries consider industrialization as an investment to accelerate their economic growth. Consequently, such a project with its modern technology and chances for employment corresponds perfectly with their intentions.

It is obvious that there exists a tendency to further process a part of the raw products within the country so as to enhance the establishment of forward and backward linkages for employment. For the investor, the climate of investment has to take into account the opportunity of direct financial support by the government (e.g., to establish the infrastructure), possibility of low taxation, additional fees, limitations for the transfer of profit, and guarantees of investment. Few developing countries are known to have laws, regulations and standards concerning environmental control. Of course, the question remains as to what extent this situation might continue within the coming 20 or 30 years.

The shortage of power supply is a worldwide problem and is shared by the industrialized as well as developing countries. Latin American countries, however, have a considerable potential with respect to both hydrocarbons and hydro-electric resources which at some places could be developed. All this requires detailed planning and time consuming negotiations with authorities.

Foremost among these concerns is a central problem, for which each country has to find its individual answer; bearing in mind its own clearly perceived interests in and need for supplies of raw materials and for measures to safeguard this supply, each country should decide within the framework of its long-range policy whether it wants to enter this new technological field of ocean mining.

CHAPTER XVI

PROBLEMS AND PROSPECTS OF TECHNOLOGICAL COOPERATION IN DEEP-SEA MINING AS THEY RELATE TO TRANSFER OF MARINE TECHNOLOGY

1. Introduction

Transfer of technology is one of the topics that have been dealt with repeatedly during the past few years at the international level. However, because of controversial opinions and differences in perception, a consensus in the form of a binding code has not been reached.

From the point of view of developing countries, transfer and practical application of technological knowledge is considered as one of the essential instruments; on one hand, it offers promise of narrowing the gap in levels of industrialization while, on the other, it promotes socio-economic and technical development in developing countries on a long-term basis. In particular, transfer of technology is aimed at the following goals: growth of employment and purchasing power; growth of output and more effective production; a more even distribution of income; and fulfillment of basic needs of the population. It is understandable, therefore, that transfer of technology has become part of the negotiations concerning the new Law of the Sea.

Transfer of technology is a very complex concept that has to be defined operationally and has to be adapted to the particular field of application. Transfer of technology is to be understood here as an operation by which technological knowledge of systems, processes and technical equipment is transmitted with the aim of practical applicability by the technology-recipient. In addition to a command of technology use the ability to develop technology further is a necessary factor. Otherwise, transfer cannot be considered as complete; the recipient can still remain in a disadvantageous position vis-a-vis the holder of the know-how.

In practice, certain modalities have evolved for transfer of technology within the frontiers of countries and beyond; technology is traded as a commodity. Its degree of innovation and its present value are indicators of its worth. In case of different technological levels between the donor and the recipient, transfer is accomplished by cooperation, forms of which vary according to the contents and the legal aspects. For example, cooperation can be directed towards the establishment of joint-ventures or can be motivated by varied interests as in the case of know-how contracts. Practical experience points to the conclusion that cooperation can be achieved successfully and on a long-term basis only when a balance of interests is reached. Consideration of essential needs and capabilities of both the donor and the recipient must be seen as the decisive factor for each cooperative effort. Exploitation of weak positions or abuse of strong positions can lead to short-term success; however, it lays the basis, at the same time, for weaknesses in the cooperative effort.

We can now discuss the question whether these issues and experiences, albeit with certain specific alterations, are applicable to cooperation in the marine field and to the transfer of marine technologies.

2. Cooperation and Transfer of Technology in Relation to Ocean Mining

2.1 Framework for Transfer the Law of the Sea Negotiation

An analysis of the endeavours to date at the international level with a view of establishing a new Law of the Sea and an examination of the existing concepts and programmes shows the great significance attributed to the subject of transfer of marine technology. Considerable efforts have been made, especially in the First and Third Committees of the United Nations Third Conference on the Law of the Sea, to formulate the general framework within which transfer of marine technology should take place. One of the basic objectives the new law is purported to achieve is the participation of all nations in ocean mining and the use of the 'common heritage of mankind' for the benefit of everyone. It is obvious from the current papers that this objective can be achieved only when transfer of marine technology is accomplished in practice; hence, marine technology transfer acquires a key role.

As of now, we have a general framework for transfer of technology which contains, essentially, the following objectives: development of suitable marine technology; collection, evaluation and dissemination of data and knowledge on marine technology; and harnessing of human resources, in particular those of developing countries, by means of technology transfer. These objectives are to be achieved, by among other means: training and education of nationals of all countries; programmes for technical cooperation at the international, regional and bilateral levels under equitable and reasonable conditions; meaningful coordination of these programmes and all assistance measures; establishment of information and other experts. The institutional framework includes specifications at three levels: the 'Authority' will be at the centre with a world-wide function of documentation, distribution and coordination; regional marine scientific and technological institutions are to be involved in the network. Transfer of marine technologies will involve two fields, namely, research and commercial use of the oceans. Within these fields the following types of functions are included: exploration; exploitation; conservation and management of marine resources; and preservation of the marine environment. International organizations, both existing and those yet to be founded, other institutions already working in this field and, above all, the governments of the individual states are requested to promote and practise transfer of marine technology.

2.2 Evaluation of Major Trends

An analysis of the current proposals, as set forth in the Informal Composite Negotiating Text[1] of July 1977, has to take into consideration the following aspects:

In these proposals, what are the basic ideas that represent a consensus? Or, in other words, what common interpretations emerge from the text itself when one abstracts from diverse points of view and different opinions expressed in various publications and papers?

Does the proposed framework include sufficient conditions for transfer of marine technologies so as to do justice to the formulated objectives?

Use of the resources of the oceans should serve the friendly relations among all states, and all nations, in particular developing countries, ought to profit from it in order to accelerate their socio-economic development. This is the basic idea which has been the foundation of all the papers and discussion in the Conference. In this context, it was recognized that transfer of technology is one of the essential means by which this objective can be achieved. This concerns all nations and one would hope that they will undertake any measures they can afford so that future development occurs in the spirit mentioned above.

Since the framework takes into account clearly expressed postulates involving mutually esteemed balance of interests as the principle guiding all acitivities, it can be considered that an appropriate basis for international cooperation is included in the framework. The aim to realize transfer of marine technology under fair and reasonable conditions and by means of generally accepted guidelines, criteria and standards which are still to be formulated is implicit in this principle.

The suggested framework can be viewed as including sufficient conditions for transfer of marine technology, and in the next stage, the transfer process depends decisively on forthcoming activities, i.e., on the extent and the kind of concrete rules and regulations emerging from further negotiations that will be incorporated in the existing model.

The future will see how far the general objectives will be incorporated into the negotiations and the extent of attention paid to them. Undoubtedly, the factor of greatest importance will be whether the negotiating parties demonstrate an ability to take into consideration the interests of one another. Cooperation requires not only objectivity regarding the ends and means of negotiations, but also the removal of resentments on either side. A joint basis has to be found and the readiness for compromise will be the decisive factor as to whether the basic guidelines will be reduced to useless declarations or will be converted into the category of a recognized code for an efficient transfer of technology supported by a cooperative spirit for the benefit of all partners involved.

If the use of marine resources is not to be regarded as a privilege of a few nations, but as a common privilege of all nations, efforts have to be made, in all possible ways, to develop appropriate technologies. These efforts should be initiated without any delay so that advantage can be taken of the relatively favourable present conditions for the purpose of further negotiations.

Following these general and basic considerations, some specific aspects of tranfer of technology in relation to ocean mining should now be considered. Furthermore, the limits of transfer and the general circumstances and conditions which make cooperation viable should also be pointed out.

2.3 Cooperation as a Prerequisite for an Efficient Transfer of Technology

As stated before, transfer of technology means that the recipient shall be able to obtain and command knowledge, use it for commercial purposes and promote its development by himself. This can only be achieved through cooperation. The advancement of the recipient can be achieved only through 'on the job training' or 'learning by doing'. Mastering the technology and the capability of pushing its development forward is indispensable for strengthening one's position and for survival in the world market.

In the discussion up to this point, the basic idea of cooperation has been taken for granted, but a relevant question still remains to be answered; how to achieve cooperation at the same time taking into consideration balanced interests, and what steps to be undertaken in order to remove existing obstacles. Negotiations are to be carried out concerning a wide complex of interests, and it is important to define quite frankly and precisely what potential partners have in mind when talking about a 'fair balance of interests'.

2.4 Technological Information and Transfer of Technology

From an analytical point of view an essential distinction has to be made between technology transfer and 'transfer of technological information'. The latter facilitates the access to technology through documentation on materials, technical processes, etc., and is a necessary step, but is quite obviously not synonymous with technology transfer.

Collection of data by the 'Authority' and by regional marine scientific and technological centres on marine equipment, as well as on 'know-how' available in written form, can be considered as appropriate steps forward in the sense that it helps establish a world-wide information network. It seems very unlikely, however, that an enterprise will be capable of undertaking ocean mining on the basis of this information, even if all the machinery, equipment and data are available. Possession of knowledge about machinery and the way it works, and the operation of this machinery in a chain of processes are two things of quite different nature. Transfer of technology is basically transmission of know-how about the latter and definitely different from export of blue prints. Only experts, as a matter of fact a team of experts, are in possession of that actual know-how. the integration of such a team, matching representatives of one party with those of counterparts, makes it possible to transmit the variety of experience resulting in a command of the processes.

This is not to say that it is, in principle, impossible to carry out processes by the do-it-yourself method; but this method involves so many difficulties and inner frictions, especially in view of the present speed and the variety of technological development, that the viability of this method is very uncertain, to say the least.

2.5 Transfer of Technology in the Research and Development Phase

The starting point, and hence the possibilities of, and requirements for, transfer of marine technology, are quite different from those for conventional technology.

In the case of transfer of technology required for building a petro-chemical complex in a developing country, the technologies concerned are those which have been proven and found economically reasonable over the years and on different occasions. In the case of ocean mining technology, not only is there a lack of past experience, but the following point which differentiates it from traditional cases has to be borne in mind also; a system of technologies for exploration, mining, lifting to the water surface and processing of nodules does not exist. Furthermore, there is no indication in what way the partial processes will be combined in order to achieve the highest economical efficiency.

Thus, it has to be pointed out that the traditional donor-receiver arrangement cannot be taken as a basis in this connection. But on the other hand, this situation gives rise to a special opportunity for developing countries to join the groups which have already solved some of the problems and have pushed forward ocean mining research. These groups have just started a long journey, and it may still be possible that developing countries are in a position to join them and to make the task facing mankind a common one.

2.6 Relationship with the Rate of Technological Development

In transfer of technology, knowledge related to the present state-of-the-art is transmitted. Innovations and development of technologies give almost hand-in-hand, particularly in those fields, involving highly sophisticated technologies such as ocean mining. This is true about both products and processes. In the electronics industry, studies show that roughly 60-80% of the products which are on the market today are less than five years old. This high rate of product and process development normally begins in the second or the third generation, while the rate of development before that stage is much slower.

This example points to the fact that the process of technology transfer cannot be fixed to a static technology of a particular moment; the transfer process instead has to parallel the dynamics of development and improvement over the period of time concerned. This is a fact which calls for an early start of the transfer process for developing nations; in the initial phases of research, the entry by a new country into a new field like ocean mining may be much easier because the lead of the innovating countries is still comparatively narrow.

2.7 The Need for Cooperation

Development of technology involves long periods of time and large amounts of investment for equipment and personnel. Companies bear these expenses as necessary ones, in order to be in a position to produce at lower cost in the long run, when the improvement of existing processes can be financed from higher profits.

Ocean mining presents quite a different picture. There are no existing processes or activities which can 'pay' for the research and development of today; perhaps there will be some in the next few years. Therefore, the people responsible for the soundness of the company have to keep the costs of research and development activities within certain limits; otherwise, the company will be exposed to very high risks. Furthermore, these limits are influenced to a certain degree by the probable opportunities of utilizing the research activities for commercial purposes later on. It is indispensable to take all these facts into consideration in order to carry out objective analysis.

Research and development of ocean mining technologies have already reached limits set from the financial point of view. The existing situation reflects this quite clearly through the fact that no single company has the capability to cover the costs involved on its own. At this stage it can be foreseen that the next financial obstacle will arise during the further phases involving pilot plants and prototypes. So far it has not been possible to work out a calculation of the forthcoming costs. There are only rough estimates, based on feasibility studies of a manganese nodule project, in the amount of about 100 million dollars.

The consortia existing today are formed by groups of industrialized countries. The question must be asked whether this situation is in conformity with the primary objective of the utilization of the 'common heritage of mankind' by means of common development of technology.

2.8 Problems of World-wide Transfer

There are innumerable practical problems in the world-wide transfer of marine technology for the benefit of all nations in the context of the 'common heritage of mankind'. These involve heavy tasks of a technical, administrative and economic nature. These tasks become more aggravating with time. It can, therefore, be safely concluded that a world-wide transfer of ocean mining technology will not be possible in the very near future.

3. Ways and Means of Cooperation

3.1 Cooperation in Scientific and Industrial Fields

The present situation, which is marked by a great variety of interests and levels of technological knowledge, excludes any simple solution. The many-sided problem can only be solved when the set of basic preconditions defined at the initial stage are observed carefully. An attempt is made in the following sections to put forward some starting points as solutions for a basis for an exchange of views. The special problems of transfer of marine technology to developing countries must be carefully kept in mind.

The possible transfer of marine technologies have to be judged by their purpose. One can ask the question; is this transfer of technology in the field of marine scientific investigation or in the field of research and development with commercial interest?

Transfer of technology in the former field does not pose any special problem, because the results of scientific investigations are available through publications anyway, and cooperation in scientific fields does not usually run into conflict of interests. Cooperation in this area is handicapped mostly by organizational and financial problems. Besides, transfer of results of scientific investigations is already being carried out successfully in many other scientific disciplines and marine science does not call for any new ways or methods of transfer.

A different situation arises when we move from the scientific field to an area where the party supplying the technology and the party receiving it are private or state commercial enterprises. In seeking solutions to the problems involved in this area the following questions have to be answered in the first place: who are the partners; through what means can the partners be brought together; in what phases does cooperation take place; through what means can the transfer of marine technology be financed.

3.2 Models for "Partnership Programmes"

The preliminary remarks have clearly shown that transfer of technology as defined in this paper can be efficient only through 'partnership programmes'. Such 'partnership programmes' can be carried out in principle as 'joint ventures' or in the form of 'know-how transfer' or as some variant of these two forms. Applied to deep-sea mining the following possibilities for these two different models are practicable:

Joint Venture: A small group of developing countries join with a consortium on the basis of a long-term contract for cooperation. Cooperative arrangements can be made covering any or all the phases; examination of feasibility; development of equipment for different components of deep-sea mining; tests of prototypes; (pre-commercial phase; and the commercial exploitation phase). In general it is essential for the operation of this model that, whatever phase is covered by the arrangement, the work is carried out as a common project with equitable distribution of gains and losses.

Know-how Transfer: A group of developing countries, represented by state or private enterprises, join and obtain the know-how by grouping together qualified experts and personnel in their project. These persons can be recruited from the free market or from associations working in the field of deep-sea mining through a 'leasing contract'. Such a group may be contracted for the whole deep-sea mining project or for individual phases of the project.

Transfer of technology, as stated earlier is, first of all, transfer of knowledge in 'human brains'; i.e. the direct transfer by the expert himself. In addition to this, appropriate equipment needs to be transferred. This equipment is no doubt, available individually on the market, but the integration of the whole set of equipment into one working unit, for instance, for the purpose of prospecting or mining, requires an enormous financial capacity. Here we should consider, for the purpose of transfer of marine technologies, the possibility of equipment-leasing or equipment-sharing. This leads to considerable savings.

3.3 Possibilities of Financing

Discussions on transfer of marine technologies always concentrate on the problem, "by whom and in what way" the financial means that are necessary for the task should be procured. Answers to this question will reflect different interests and opinions.

In the long run, a solution can be expected when ocean mining enters into the commercial phase. But then the basic question arises; is it meaningful and practicable if developing countries enter into the process of transfer of technology at a late stage? The gap between those who possess technology and those who are in need of it is steadily increasing, and, following this trend, the position of the potential technology-recipient is increasingly weakened with the passage of time. At present, marine research activities for deep-sea mining have been going on only for a few years; the gap between technology-developer and technology-recipient is not insurmountable and it seems possible that the existing gap can be eliminated by appropriate forms and modes of cooperation.

The question concerning financial arrangements has to be answered in the context of this background. Furthermore, an analysis of the circumstances has to be carried out critically. It is not the industrialized countries, but the private industry of these countries that possesses the technology and the know-how. These companies are clearly oriented towards commercial targets. It is unrealistic to assume that they will initiate a transfer of technology without receiving any equivalent return. Exploration activities which take place on the continental shelf and within the economic zone are subject to certain limitations as some developing countries do not wish that their resources be exploited by others and exported for others' use.

Taking these issues into account the following two possibilities seem worth examination: First, financing of marine technology transfer by existing international organizations. It has to be ascertained whether or to what extent the programmes of these organizations are geared to marine technology transfer and what modifications are necessary. The use of international organizations will offer one great advantage; since they have a proven system of performance and control available. Also, they are in a position to make a competent assessment of the complexity of financial concerns relating to the project. Secondly, financing of marine technology transfer through bilateral contracts on techno-economic cooperation. Ocean mining, as a unit, can be incorporated in these existing contracts. An exploration in the 200-mile economic zone as well as activities in the deep-sea can be the basis of a future contract. In the latter case it seems likely that transfer of technology will be financed partially by the resources of the recipient country.

There is another arrangement to which attention should be drawn, namely a triangular arrangement consisting of three parties: one party involved in ocean mining; the second one a country with raw material resources; exports and surpluses thereof; and the third one a developing country. A modified form of the 'partnership programme' can be worked out.

These possibilities should be thoroughly examined and discussed at the international level.

4. Conclusions

Transfer of marine technologies should not be considered as a problem belonging to the future. It is definitely a problem, the initial steps for the solution of which have to be taken today.

If the nations of the world are really interested in measures which will, firstly, halt the widening of the gap between the industrialized countries and the countries of the Third World and, secondly, narrow the gap, mankind, then, should take advantage of this new scientific-technological opportunity which is capable of harnessing human as well as natural resources.

PART VII

CONCLUSIONS

The Group of Experts heard a series of papers presented by members, questioned the authors of these papers, tapped additional information from members with further specialized knowledge, and evaluated through open discussion their common understandings with respect to assessment of sea-bed mineral resources. This chapter summarizes the basic findings and conclusions of the Group of Experts.

The approach of the Group of Experts to the task of assessing sea-bed mineral resources was to isolate the various factors of importance, while recognizing their complex interrrelationship, and then to attempt to analyze these isolated factors. Finally, the factors were recombined to reach a common basis for assessment of the sea-bed mineral resources.

1. Resources In-Situ

In evaluating manganese nodule resources in-situ, the following distinctions were agreed upon. First, the term "in-situ" as used, referred to manganese nodules lying undisturbed on the sea-bed with no factors of recoverability taken into account. Second, "manganese nodule resource" was used to denote manganese nodule deposits that were likely to be economically workable at some time in the future and that included both known deposits and deposits whose existence could only be deduced.

It was not possible to characterize any deposit as a mineral reserve since none was being economically worked at present. Some part of the resource might be described as a potential reserve on the basis that it would be workable by the first mining units.

The question of what would be economically workable deposits at some time in the future could not be answered in a precise manner because of the many basic economic uncertainties. However, the Group of Experts agreed that such deposits would have to be characterized by a minimum average grade (content of nickel plus copper) and a minimum average abundance (weight per unit area) on the sea-floor. As in most ore bodies, the metal content was the more important of these two factors.

In the absence of criteria for defining 'resource' in terms of grade, the consensus of the meeting was that at the present time, it was reasonable to assume that the lowest average grade required to qualify as a resource was about 1.5% combined nickel and copper. The minimum acceptable average grade required to qualify as a potential reserve was about 2.25% combined nickel and copper. It was noted that the ratio of nickel to copper in these combined grades was about 5 to 4.

Similarly, the lowest acceptable average abundance required to qualify as a resource was about 5 kg/m^2 (wet), and to qualify as a potential reserve, about 10 kg/m^2 (wet). Possible combinations of grade and abundance that led to a higher in-situ weight of potentially recoverable metal per unit area could increase the total amount of both resources and potential reserves.

Estimates of in-situ quantities varied widely because of the inadequacy of the data with respect to reliability and amount. The estimation of the tonnage of manganese nodules meeting the agreed criteria necessitated data from a sufficient number of reliable and evenly-distributed sample stations. The existing information available in the public domain did not meet these requirements. According to the information in the data bank of the Scripps Institute of Oceanography, about 55,000 locations on the seafloor had been sampled or photographed. Of these, about 3,800 analyses of manganese nodules had been made on samples from 1850 locations. There was, therefore, only sparse coverage of sample stations (roughly, an average of one sample station in 35,000 sq. km. of the total area in which nodules might be present.) In fact, of the above 1,850 locations, about one half were in the North Pacific, a quarter in the South Pacific, and the remainder divided between the Atlantic and the Indian Oceans. In the North Pacific, only the Clarion-Clipperton zone (defined as extending approximately from 7° N to 15° N and from 120° W to 150° W) had been sampled sufficiently to identify it with confidence as a prime area of nodules. Even there, the density of analyzed samples was about one station in 15,000 sq. km. and it was impossible to estimate either resources or potential reserves with any probable accuracy. Outside the Clarion-Clipperton zone, the data was so sparse that it was possible only to identify potential regions for prospecting and exploration. Better estimates could be made only by obtaining more data on grade and abundance. However, the rate of the accumulation of such data was low, and many years would pass before there could be a meaningful assessment outside the Clarion-Clipperton region. Nevertheless, the data could be improved by concentrating sampling in the promising areas that had been identified. While it was true that a large amount of data existed in private hands, it pertained primarily to the Clarion-Clipperton region. Thus, this data could only improve confidence in the Clarion-Clipperton estimates and could not materially improve the accuracy of world-wide estimates.

2. Data on Grade and Abundance

It was generally acknowledged that information presently in the public domain was neither sufficiently accurate nor sufficiently abundant to permit a precise determination of the resources and reserves beyond first order resource inventory studies. For this a large, publicly accessible world-wide data base on nodule grade and abundance was essential but would take many years to develop. Future work, therefore, should be directed to increased sampling coupled with the gathering of information on micro and macrotopography both in areas which were currently sparsely sampled but showed relatively attractive grades, and in new areas. At the same time, there was a need to ensure that the data collected were as accurate and representative as possible. To that end, standardization of sampling techniques and of methods of analysis was highly desirable. Exchange of data and cooperation among various laboratories should be encouraged.

Improving reserve estimates also was inevitably tied to the improvement of the estimates of operating costs and must therefore await, *inter alia,* the outcome of full scale testing particularly of the mining subsystem. Finally, improvements in methodology that allowed uncertainty to be dealt with could probably advance resource estimates a step further.

3. Technical Characteristics

The basic technology for mining and metallurgical recovery was available. Ongoing commercial research and development programmes were largely centred on predicting engineering performance or operability. The determination of areal requirements depended on such factors as water depth, topography, grade, abundance, dredge efficiency, sweep efficiency, manoeuvrability of the mining ship or ships and metallurgical process recovery. These factors, however, were site specific. A mine-site must meet certain requirements, which on present evidence, might include a minimum minable portion of 60% to 90%; an average abundance between 6,500 and 10,000 mt/km^2 (wet); and an average grade (combined nickel and copper) of 2.25%[1]. A dredge efficiency of at least 30% and a sweep efficiency of at least 40% might also be assumed. It must be emphasized that these factors were interrelated and they must be taken as a whole. Taking into account these factors, it might be said that the overall efficiency of mining systems would be likely to fall in the range of 10% to 40%.

Production volumes would be determined by the economies of the total system and would involve in addition to the operating costs estimated on a proven technically reliable system, estimates of the revenue from the products, the investment required and the financing costs. It seemed clear at the present time that the probably lower limit of annual production volume from each deep-sea mining operation would be somewhere between 1,000,000 and 3,000,000 dry metric tons, with actual capacity probably dependent upon whether manganese would be recovered or not.

Whichever processing methods (pyrometallurgical or hydrometallurgical) were used to recover the metals, the consumption of energy was likely to be extremely high. There would be additional associated costs of meeting satisfactorily environmental requirements. The location of a processing plant in developing countries within the circum-Pacific region seemed to merit serious consideration.

Even though the commencement of commercial nodule mining would be determined by a number of factors, including the status of preparedness of each consortium and the existence of an acceptable legal regime, the following were suggested as general time horizons for the incipient industry from a technical standpoint:

a) Following an investment decision, as few as *three years* might be required before commercial production could commence. Four of the consortia were, or had announced that they would be conducting prototype tests in the next two years.

b) A total of about *five years* would be needed to obtain the requisite approvals and to construct a full scale process plant.

c) The earliest date for commercial production of nodules from the sea-bed would be in the early 1980's with the mid-1980's as the most probable.

In closing, it was stressed that considering the short time at their disposal and the limitations of available data, the Group of Experts had addressed, and had succeeded in clarifying the major issues associated with the assessment of sea-bed mineral resources, as well as suggesting measures for improving the accuracy of such assessments.

It was also emphasized that while the results of the meeting would make a significant contribution to ongoing efforts, sea-bed mining was still an activity of the future and as new data and information became available in this rapidly developing field, it would soon be necessary to re-examine and update the findings of the meeting.

Index of Names

Index of Subjects